RECENT TRENDS IN COMBINATORICS

The legacy of Paul Erdős

Paul Erdős: 26 March 1913 to 20 September 1996.

RECENT TRENDS IN COMBINATORICS

The legacy of Paul Erdős

Edited by

ERVIN GYŐRI
VERA T. SÓS

*Mathematical Institute of
the Hungarian Academy of Sciences*

CAMBRIDGE
UNIVERSITY PRESS

CAMBRIDGE UNIVERSITY PRESS
Cambridge, New York, Melbourne, Madrid, Cape Town, Singapore, São Paulo, Delhi

Cambridge University Press
The Edinburgh Building, Cambridge CB2 8RU, UK

Published in the United States of America by Cambridge University Press, New York

www.cambridge.org
Information on this title: www.cambridge.org/9780521120043

First published 2001
This digitally printed version 2009

A catalogue record for this publication is available from the British Library

ISBN 978-0-521-80170-6 hardback
ISBN 978-0-521-12004-3 paperback

Contents

Preface

The combinatorial workshop 'Some Trends in Discrete Mathematics' was held in Mátraháza, Hungary, from 22 to 28 October 1995. The aim of the workshop was to expose connections between distant parts of combinatorial mathematics, such as pure combinatorics, graph theory, combinatorial number theory and random graphs, by bringing together researchers from diverse fields. To emphasize the workshop character of this meeting, we invited many distinguished mathematicians but asked only ten of them to give lectures. (Unfortunately, illness prevented Claude Berge from attending the meeting.) There were no contributed talks, but the lectures were followed by long discussions involving all the participants: these sessions played a crucial role in the success of the workshop. A tangible result of these evening discussions is the Cameron–Erdős paper in this volume.

A highlight of the volume is the paper Paul Erdős was writing on the eve of his sudden death in Warsaw on 20 September, 1996. This paper had no title and, except for light editing, this very special manuscript is published as he left it. The other eight papers of this issue are surveys and research papers written by the invited speakers and their collaborators.

We want to thank all the participants of the workshop for their contribution to its success. We are also grateful to DIMANET and its main coordinator, Professor Walter Deuber, for providing the financial support that made the workshop possible. We wish to express our sincere thanks to Béla Bollobás who made it possible to publish these papers in a special issue of *Combinatorics, Probability and Computing*.

Ervin Győri and Vera T. Sós

I will try to present either new problems or old ones which in my opinion have been underservedly forgotten or neglected. The problems and results will all have a combinatorial ~~flavor~~ flavor.

1. This problem was started by Hajnal and myself. First a few words on set theory. Let S be a set, to every proper subset S' of S we make correspond an element x of S $x \cap S' = \emptyset$. $f(S') = x$. A subset S' of S is called independent or free if for every subset S'' of S' $f(S'') \cap S'$ is empty. In our first paper with Hajnal we proved that if $|S| < \aleph_a$ we can always find a $f(S')$ so that there should not be an infinite independent set. If $|S| = \aleph_a$ we could not decide if there always is an infinite independent set. This problem is perhaps undecidable.

In a more recent paper Hajnal and I investigated the following finite problem. Let $|S| = m < \aleph_0$ $h(m)$ is the largest integer for which for every f there is an independent subset $S_1 \subset S$, $f S_1 |S_1| = h(m)$. Further let $~~Fast~~ H(m)$ be the smallest integer for which there is a function f for which for every $S'' \subset S$, $|S''| \geq H(m)$, $f(S') = S$ where $F(S'')$ is the union of all the element $f(S'') \subset S''$.

A facsimile page of the first draft of the paper of Professor Erdős in this volume.

Foreword
Paul Erdős: The Man and the Mathematician (1913–1996)

MIKLÓS SIMONOVITS and VERA T. SÓS

Mathematical Institute of the Hungarian Academy of Sciences,
13–15 Reáltanoda utca, 1053 Budapest, Hungary

1. Preface

Paul Erdős was one of the greatest mathematical figures of the twentieth century. An enormous number of obituaries have appeared since his death, which is very unusual in the world of science.[1]

Paul Erdős had already become a legend in his own lifetime. With a strong character, a clear moral compass and an incredible love of mathematics he created a new mould for a life style. (Will anyone else ever fit it?) This did not require him to pay attention to some of the details of living that most of us deal with. Thus it is not surprising that he developed the eccentricities which are often related by people who knew him and in articles about him – sometimes accurate and sometimes exaggerated. Some of these stories he did not care for, but others he liked to remember, and would retell himself, contributing to the canonization of these anecdotes. As with other passionate geniuses, stories about his eccentricities are a way for writers to show how unusual he was. However, to those who knew him closely, these stories, although amusing, do not in themselves capture the essence of this person, who was so very connected to the world.

Here we shall try to depict Erdős as *we* saw him.

1.1. Mátraháza, 1995

This volume is the Proceedings of the Mátraháza Workshop, one of those workshops which he liked so much, where he felt at home, where he was surrounded by old friends and young mathematicians, all eager to speak with him, to ask him questions, to tell him their results.

There was something special about Mátraháza, a small village 100 km from Budapest. It is a tourist resort on the Mátra mountain, which is the highest mountain in Hungary,

---

[1] Among other places, one could read about him in the *New York Times*, the *Washington Post*, the (London) *Times*, and even in the latest edition of the *Encyclopaedia Britannica*.

although only about 1000 metres high. The Hungarian Academy of Sciences has a nice holiday home there for researchers and their families.

Erdős liked to come here. In the late 1960s and early 70s, Erdős often spent a week or two here, with his mother and his friends, like the Kalmárs, the Rényis and the Turáns,[2] or with others, mathematicians and non-mathematicians, working, walking and discussing politics, history, economics, often with some of the best experts in Hungary. This was also the place where he played much table tennis (in his funny but very effective way), chess and go. (For a long time he was one of the best Hungarian go players.) Some of our favourite papers were born here, e.g. [36], [37].

Erdős had his own favourite trails here, one of them leading to a small tower which he liked to climb. Yes, Erdős liked this place and, above all, he liked to be here with his friends.

In 1995 we organized a one week workshop at Mátraháza for a small group of people in combinatorics and number theory. We spent a week there "proving and conjecturing". We remember Erdős working hard with Peter Cameron on some problems in combinatorial number theory. The theorems proved there in the evenings or late nights, while a few metres away other people were discussing completely different mathematical problems, can be read in this volume. Their paper was partly born in front of us. Erdős was surrounded here with love, reverence and care, and there was a warm and friendly atmosphere.

1.2. Warsaw, one year later

Erdős died in Warsaw, on 20 September, 1996. The circumstances of his death were profoundly moving. There was a five week long Mini Semester in Combinatorics at the Banach Center, in Warsaw. He participated in this series of workshops for two weeks. He enjoyed the mathematical atmosphere, but at the same time he often complained about the cold weather. He gave two lectures, the second one on Wednesday, 18 September, receiving a big round of applause. Early on Saturday morning he was taken to hospital from his hotel room, following a heart attack. It is very likely that he soon lost his ability to communicate and then he lost consciousness. Around 3 pm he had a second, much more serious, heart attack and this one killed him.

What we all feel was expressed by Vojta Rödl:

"Things won't be the same without Uncle Paul".

* * *

Wherever he travelled in the world, during almost every moment of the day, he was surrounded by colleagues and friends who cared about him. This was something which meant more and more to him as he grew older. Unfortunately, his last hours were spent in an unfamiliar hospital, alone.

---

[2] László Kalmár helped Erdős to write his first paper, a new, simplified proof of Chebyshev's theorem on prime numbers.

Erdős often explained that the most beautiful way to "finish life" is "... to give a lecture, finish a proof, put down the chalk and die...". In some sense he passed away as he wished. His brain was engaged in mathematics until the very last day. This is what he wished so much all the time.

2. The travelling mathematician; the co-author mathematician

In 1939, for some strange reason, Erdős's fellowship at the Institute for Advanced Studies in Princeton was not renewed. It was at that point that he started his "nomadic" life, travelling from place to place. Nobody can tell what kind of life he would have had if his fellowship at the Institute had been extended.[3]

The important part of his life style was not just that he visited many places. It was much more important that he was able to make "human" and "mathematical" friendships, very easily, wherever he went. Arriving at a new place he would meet several mathematicians, ask them about their research, start thinking about their questions, and quite often would give striking solutions to their mathematical problems. Needless to say, he was incredibly fast.

It often happens that someone so powerful and deep mathematically, and so fast-thinking, can be intimidating to work with. It can be rather frustrating if one thinks about a problem for months and somebody just comes around, asks a few questions, and then solves it. To work with Erdős was completely different. Whenever you spoke with him, even though you knew that a mathematical giant was sitting there, opposite you, he could still create a very special atmosphere, where you felt equal. He was always asking questions about problems near to your mathematical interests, or just questions where he got stuck. Everybody was happy to work on his questions, and felt proud if he succeeded in solving an "Erdős problem" and so had a chance to explain the solution to him. But he would not stop there: he went on by asking further related questions, and never stopped pursuing new directions connected to the original innocent-looking problem.

We all have stories about this. Let us illustrate with a story of András Hajnal. (Hajnal describes this whole story in more detail in [42].) Hajnal was a PhD student of László Kalmár, in Szeged, working in set theory. Erdős met Hajnal while visiting Kalmár. Erdős asked Hajnal what was he doing mathematically, and Hajnal explained his topic. Erdős listened politely to Hajnal (in spite of the fact that he was not interested much in that particular part of mathematics) and then, suddenly, switched to some other topic, by asking: "... and are you interested in normal set theory as well?" With that, they started discussing mathematics which was interesting for both of them. Later Erdős and Hajnal climbed the tower of the Szeged Memorial Church. (As mentioned above, Erdős very much liked climbing hills, mountains, towers.) Hajnal writes:

> "... I had by then lived for two years in Szeged, and I had never had the slightest difficulty in resisting any pressure to visit the tower. However, to my surprise, I could

---

[3] L. Babai [2] writes in his very detailed and informative article about Erdős: "To his [Erdős's] dismay, his IAS fellowship is not extended and he is left penniless. 'Leopold Infeld wasn't extended either,' he comments."

not resist this invitation. Climbing those stairs more results and conjectures were formulated by him [Erdős], while, at the same time, he was complaining that he felt a little dizzy.

That day ended with dinner at Kalmár's house where the conversation continued mainly about set mappings but was sometimes interrupted with some of his comments on "Sam and Joe" [the USA and the USSR]. When we parted, it was almost as from an old friend – there was a joint-paper half ready which could be completed by correspondence."

That day completely changed Hajnal's life.

* * *

Erdős helped people in various ways: mathematically, or writing letters of recommendation, or lending money. He kept visiting his old friends just to "keep them company". For emotional reasons, he stopped using his apartment in Budapest after his mother's death (instead, he lived at the guest house of the academy), and for many years his friends in need lived there: some Hungarians, while changing from one apartment to another, and many of his friends from abroad, while visiting Hungary. Erdős cared for people. He was a sympathetic and compassionate person. It really made him feel depressed if one of his friends got into a difficult situation.

He often was invited to his friends' houses to stay with them. The following poetic lines, written by Paul to the second author in 1976, could have been written from many places.

"It is six in the morning. The house still sleeps. I am listening to lovely music; I write and conjecture."[4]

* * *

Friendship was very important to Paul. When Erdős became a university student, he met Tibor Gallai, Géza Grünwald, Eszter Klein, Pál Turán, Endre Vázsonyi and some others from the Pázmány Péter University, Budapest,[5] and also György Szekeres, a student in chemical engineering at the Technical University, who was shuttling between these two universities to attend classes in both. This was the start of several life-long collaborations and friendships, some tragically cut short by the coming Fascism, increasing antisemitism and then by the Second World War. Life became more and more difficult in Hungary. Erdős was among those who soon realized that Hungary was no longer safe for Jews. So he left.

Remarkably, these people, with the exception of Géza Grünwald, survived the holocaust, but many of their friends and relatives were victims of Fascism.

Let us return to the university years. According to their reminiscences, their regular meetings, which were day-long excursions around Budapest or in the City Park, had

[4] In Hungarian: "Reggel hat van s a ház még alszik – szép zenét hallgatok, s közben irok és sejtek."
[5] Now called Eötvös Loránd University.

a great impact on their future lives in all respects. It was a kind of open, peripatetic university, brought about by the socio-political discrimination of the era which affected most of them.

This is what Szekeres wrote about these years: "We had a very close circle of young mathematicians, foremost among them Erdős, Turán and Gallai; friendships were forged which became the most lasting that I have ever known and which outlived the upheavals of the thirties, a vicious world war and our scattering to the four corners of the world. Our discussions centered around mathematics, personal gossip and politics." (Later Szekeres and his wife also left Hungary.)

The relationship of Erdős to his mother was legendary. Because of the war and the political situation he could not see his mother for ten years. After 1955 he visited Hungary regularly and his relationship with her changed: they spent more and more time together and this became one of the dominating features of his life. His mother's death shocked and changed him completely.

3. *Ars mathematica*: To conjecture and prove …

The best way to learn about Erdős' mathematics is to read some of his original papers. Some of these, primarily in combinatorics, but also in number theory and geometry, are collected in *The Art of Counting, Selected Writings of Paul Erdős* [12].

Several survey papers on his work appeared on the occasion of his 80th birthday or since his death, and several further ones are planned for publication. Here we mention only a few: Bollobás [4], [5], Hajnal [42], and Simonovits [45]. Further, we warmly recommend the paper of Turán [52], a paper of Erdős himself [27], and also the long paper of Babai [2] and the paper of Sós [47]. We should also mention the book of Chung and Graham on problems of Erdős in graph theory [7].

Finally, we can learn a lot about Erdős' mathematics and personality by reading *his* papers about his friends: Turán, [14], [15], [16], [17], Gallai [20], Kalmár, [13], Gödel [24], Gabriel Dirac, [18], Ernst Straus [21], Ulam, [22], Richard Rado [23], to name only a few.

Erdős was a mathematician primarily interested in "concrete problems". Many mathematicians prefer to build theories, and when they get stuck they look at special cases to try to clarify the details.

Erdős used the opposite method. He set out from subcases, attacked them, proved more and more results around these particular problems, and thus built up first a small kernel of what later developed into a whole theory. For example, trying to answer some problems of Sidon he started using random methods in number theory. At the same time, he observed that some number theoretical functions behave very much like sums of independent random variables: with Kac [30], [31] and Wintner [40], [41] he started a whole new branch of number theory. This work was continued by many others, for example Kubilius, and is now described in several monographs, including the book of Elliott [8].

Erdős and Szekeres started from a "strange", seemingly tiny problem of Eszter Klein:

Given n points in general position in the plane, for how large a k can a convex k-gon always be selected from the set? One can easily see that out of five points one can always choose four in a convex position. But to prove that out of any nine points one can select five points in a convex position is much harder. Erdős and Szekeres proved a general theorem. To prove this theorem, in fact, they rediscovered Ramsey's theorem [39], which had been proved by Ramsey somewhat earlier.[6]

Answering a Ramsey question of Turán, Erdős started using random methods in graph theory.[7] Later, using more and more sophisticated versions of this method, he proved the existence of graphs with high girth and high chromatic number [10], [11]. Soon Rényi joined the investigation and they wrote a whole series of papers [32], [35], [34], [33] trying to answer the question:

What is a random graph and what does it typically look like?

Today the theory of random graphs is an important branch of combinatorics.[8] The significance and impact of Erdős's results cannot be separated from his ars mathematica. The concept of uniform distribution and the related discrepancy theory play a fundamental role in several branches of mathematics (diophantine approximation, measure theory, ergodic theory, geometry, discrete geometry, numerical analysis, theoretical computer science, etc).

* * *

His friend and co-author Ernst Straus (who was also a co-author of Albert Einstein) wrote about him: "The prince of problem solvers and the absolute monarch of problem posers". Erdős describes this as follows: "Problems have always been an essential part of my mathematical life. A well chosen problem can isolate an essential difficulty in a particular area, serving as a benchmark against which progress in this area can be measured. An innocent looking problem often gives no hint as to its true nature. . . " [28]. His problems typically have a very simple formulation, and often even the experts of a field realize only years later that Erdős posed the first generic question for a basic, general theory (which can be extremely difficult to answer): see e.g. [19].

The age when mathematicians could deeply understand and influence several distant fields in mathematics seems to have passed. But Erdős was definitely a giant of this kind. He did pioneering work in several unrelated fields, often jointly, and is helped create several branches of today's mathematics. On some occasions somebody else had written one or two pioneering papers and then Erdős began asking questions, conjecturing and

---

[6] For a detailed description of the story see e.g. Szekeres' paper [48] in *The Art of Counting: Selected Writings of Paul Erdős* [12].

[7] Setting out from Ramsey's theorem, which was reinvented by Erdős and Szekeres, Turán arrived at his famous extremal graph problem [51] and then turned back to Ramsey theory, by asking: How large a complete subgraph must occur in every graph or its complementary graph? Turán thought that $\sqrt{n}$ is the correct answer, and asked Erdős, who proved instead that $\log n$ is the correct order of magnitude. In some sense, this was the birth of random graph theory. For details see [53].

[8] The interested reader may try to read the original papers of Erdős and Rényi, or the beautiful books of Erdős and Spencer [38], Bollobás [3], Alon and Spencer [1], or the survey of Karoński and Ruciński [44].

proving theorems, and the whole area changed from a collection of a few isolated results into a blossoming theory. Often his colleagues did not immediately realize what was going on: whether Erdős' questions and answers would just be the last missing pieces of a small area or whether they would develop into a substantial theory.

When Erdős turned 50, Turán wrote a survey [52] of Paul's mathematical achievements. Turán's lines are even today among the best sources for understanding Erdős' mathematics.[9] In his "introduction" Turán wrote:

"...his works published so far (!!!) relate roughly to the following topics: Number Theory, Probability Theory and Ergodic Theory, Graph Theory and Asymptotical Combinatorics, Constructive Theory of Functions, Set Theory and Set-theoretical Topology, Theory of Series, Theory of Analytic Functions, Geometry...

...But to write [about Erdős's work] has been made especially difficult by the fact that reviewing his work and influence is hard to separate from his mathematical personality. This is partially indicated by the fact that he has joint papers with more than 90 co-authors from four continents.... In a certain sense he is an occidental Ramanujan, with his strength and limits, a singular and unique phenomenon...."

On another occasion, when Erdős was 43, Turán compared Erdős's performance in mathematics to that of Mozart in music, and mentioned that Erdős was barely 20 when he was called the magician of Budapest (*der Zauberer von Budapest*) by I. Schur.

These lines were written more than 30 years ago. At that time Erdős had published roughly 400 papers; today his total publication list approaches 1600 papers. In his later years he moved away from analytic fields towards the more combinatorial parts of mathematics. It is interesting to observe the resulting changes in the number of papers and co-authors.

4. Connecting distant fields

One basic feature of Erdős's style was to ask surprising questions connecting distant mathematical areas. He did it so many times that we got used to it and often forgot to be surprised.

In the first case he forged links between number theory, combinatorics and graph theory and geometry. More than fifty years ago Paul asked the following question [9]:

"At most how many unit distances can occur among n points in the plane?"

The question may seem to concern geometry but, obviously, it bears no similarity to the usual questions in geometry.

The conjecture is that there are at most $O(n^{1+o(1)})$ unit distances among n points, and the motivation for this conjecture is that one can scale a grid so that the number of unit distances in a disk of radius $c\sqrt{n}$ is slightly more than linear. The conjecture says that this is optimal.

There is a clear connection with the following classical question in number theory:

---

[9] It is not that surprising that many of us, writing of Paul, still use Turán's lines. Perhaps it is more interesting that Erdős, while writing one of his last papers [27], also liked going back to this source.

given an integer $m > 0$, in how many ways can it be represented in the form $x^2 + y^2$, where x, y are also integers?

This problem of Erdős is extremely simple to state, but it is still unsolved. He immediately noticed that a simple extremal graph theorem concerning forbidding a two by three complete bipartite subgraph provides the estimate $O(n^{3/2})$. A much more involved argument of Józsa and Szemerédi [43] gave an upper bound $o(n^{3/2})$. Many years later Spencer, Szemerédi and Trotter [46] proved that the number of unit distances can be estimated from above by $O(n^{4/3})$, and Brass [6] has shown that this may be sharp in some Minkowski geometries. The importance of this last result is that it shows that – if the conjecture is true – one has to use the Euclidean structure of the plane in a non-trivial way.

In the last 50 years the distribution of distances (in general metric spaces as well) has become a difficult and extensively investigated topic. Rather surprisingly, these kinds of problems have also become relevant in computer science.

A similar case is connected with the celebrated theorem of Van der Waerden about arithmetic progressions. In the early forties Erdős and Turán asked: What is the maximum length of a sequence of integers in $[1, n]$ not containing a k-term arithmetic progression? They conjectured that this maximum is $o(n)$. K.F. Roth proved this in 1954 for $k = 3$, and in 1973 E. Szemerédi proved it for the general case [49] (for which he collected a $1000 prize from Erdős). H. Fürstenberg and I. Katznelson proved many generalizations of this result using tools from ergodic theory. In the last few decades a whole new area has developed around this innocent looking question.

5. Papers, lectures, letters

Papers, lectures, correspondence and personal conversations were all important means by which Erdős could communicate about mathematics. He developed a specific genre in his papers and lectures. He wrote about 200 problem-papers that formulate dozens of problems clustered around a particular topic, always providing the background and some remarks in connection with the partial results of the moment.

In recent decades his lectures, whether plenary addresses at international congresses or educational lectures for teachers and students, took on the same characteristic form: results, partial results and conjectures interspersed with some stories that became known as "Erdősisms". It was a pleasure for him to engage in a mathematical conversation, should his partner be a high-school student or a famous mathematician.

He had a legendary memory. He could recall results of his own and others with the place and date of their publications, conversations with hundreds of mathematicians several decades ago, as one recalls yesterday's events.

His correspondence, an integral part of his daily routine, should also be mentioned. Some mathematicians may have kept several hundreds, others perhaps a dozen of his letters, which he wrote in his special characteristic style, switching abruptly from old and new mathematical problems to politics, friends and colleagues, and back.

He formulated thousands of problems and described related results and passed them on in his letters, between countries, through geographical and political borders, stimulating close friends and casual acquaintances to work together. Many of his results were obtained entirely through correspondence.

6. Shifting the emphasis to combinatorics?

The first two decades of Erdős's mathematics were dominated by number theory and analysis. Nowadays we think of the majority of his work as combinatorial mathematics: graph theory, combinatorial number theory, combinatorial geometry, combinatorial set theory. As a matter of fact, his earlier work had a strong combinatorial flavour as well.

Combinatorics, in close interaction with computer science, has undergone explosive development in the past 2 or 3 decades. Though Erdős himself was never directly involved in computer science, the indirect influence of computer science can be found in his mathematics. Yet the influence of Erdős's mathematics in computer science is much stronger. For example, his "random method" became one of the most powerful tools in Theoretical Computer Science; among other things, it is used to give lower bounds on the running time of algorithms.

We asked him to write a paper for the volume which was published in honour of his 80th birthday with the title "On my favourite theorems" [27], instead of the usual "On my favourite problems". This helps us to select some of his most important results by giving us his own opinion.

Erdős found elementary proofs for many classical theorems in prime number theory. (Here, "elementary" means that it does not use complex function theory.) In 1948 Erdős and Selberg found a long-awaited elementary proof of the prime number theorem.

Erdős and Rényi founded the theory of random graphs in the early 1960s. The significance of this field keeps growing. The systematic application of random methods in number theory, graph theory, and in many other fields is one of Paul's most important contributions.

In addition to the above mentioned fields, Erdős contributed to many others from his early youth until the end. An example of such an important area is interpolation theory. It is not the aim of this paper to cover these topics. They all deserve one or more separate surveys, which will surely be written by various authors in the future.

7. Did Erdős care only for numbers?

Paul's lifestyle centred around three values – independence, search for truth, and caring humanitarianism.

To secure his personal independence (independence from the political powers, doctrines and conventions of the surrounding world), he gave up possessions, family, and a safe job. This was not possible without sacrifices, as he himself often remarked. He paid the price for it all his life. The search for truth in science, politics, and everyday life provided the motivation for his round-the-clock activity and productivity, ignoring pains, sickness and aging. He continued his nomadic life style until the very end, writing 20–30 papers and giving several dozens of lectures yearly.

Despite the opinion of many people, he was not ascetic, even though he did not value possessions. On the contrary, he enjoyed life very much. He liked nature. He liked reading books on various topics. Sometimes, upon entering one's room, he picked up a book, read a few pages from it, asked if he might borrow the book, and then sent it back by mail from some other part of the world. He was astonishingly well versed in literature, biology, history and politics. He liked music, particularly Bach, Händel and Mozart. While working on mathematical problems, he often asked us to put on records or listened to his radio.

He liked going to restaurants and had a gourmet's taste (without a gourmand's appetite). He liked to invite people to restaurants and he liked to be invited to a good party or to a good dinner. He liked to live in good hotels.

However, in India he refused to eat decent food and would not go to good restaurants. It was not for shortage of money. He thought that in a country where hundreds of millions go hungry he should not eat like a gourmet. When he delivered a lecture about child prodigies at the Tata Institute in Bombay, he remarked to his mostly well-to-do audience that he did not understand how one could accept prosperity in the midst of such omnipresent poverty.

He did not value possessions but, as a matter of fact, he did not need them either. When he wanted to leave a place, the airplane ticket was there and he just left.

* * *

Of course, there were different periods in Erdős's life. Whatever we write about him will be more characteristic of some of these and less typical of others. Yet this is not the place to go into a detailed discussion of these differences.

In spite of countless stories, many of which have been further embellished after his death, he will be remembered clearly as a brilliant, unique mathematician of pure character, and as a warm, compassionate and charitable person. He was exceptional as a mathematician, as an intellectual, and as a human being with deep feelings who carved out a remarkable life for himself. We hope that this is the picture of Paul Erdős that will emerge in time and will endure.

Acknowledgement

We would like to thank John Goldwasser for his very valuable help in improving the style of this paper.

References

[1] Alon, N. and Spencer, J. (1992) *The Probabilistic Method*, Wiley Interscience.
[2] Babai, L. (1996) In and out of Hungary, Paul Erdős, his friends and times. In *Combinatorics: Paul Erdős is Eighty*, Vol. 2/2, *Bolyai Society Math Studies*, Keszthely, Hungary, pp. 7–96.
[3] Bollobás, B. (1985) *Random Graphs*, Academic Press, London.
[4] Bollobás, B. (1997) Paul Erdős — life and work. In *The Mathematics of Paul Erdős I* (R. L. Graham and J. Nešetřil, eds), Algorithms and Combinatorics, Vol. 13, Springer Verlag, pp. 1–41.
[5] Bollobás, B. (1998) To prove and conjecture: Paul Erdős and his mathematics. *Amer. Math. Monthly* **105** 209–237.

[6] Brass, P. (1990) Erdős distance problems in normed spaces. *Comput. Geom.* **6** 195–214.

[7] Chung, F. and Graham, R. L. (1998) *Erdős on Graphs: His Legacy of Unsolved Problems*, A. K. Peters Ltd, Wellesley, MA.

[8] Elliott, P. D. T. (1980) *Probabilistic Number Theory*, Vols 1–2, Springer Verlag.

[9] Erdős, P. (1946) On sets of distances of *n* points. *Amer. Math. Monthly* **53** 248–250.

[10] Erdős, P. (1959) Graph Theory and Probability. *Canad. Journal of Math.* **11** 34–38.

[11] Erdős, P. (1961) Graph Theory and Probability, II. *Canad. Journal of Math.* **13** 346–352.

[12] Erdős, P. (1973) *The Art of Counting: Selected Writings of Paul Erdős* (Joel Spencer, ed.), MIT Press.

[13] Erdős, P. (1974) Néhány személyes és matematikai emlékem Kalmár Lászlóról (My personal and mathematical reminiscences of László Kalmár, in Hungarian). *Mat. Lapok* **25** no. 3–4, 253–255.

[14] Erdős, P. (1977) Paul Turán 1910–1976: His work in graph theory. *J. Graph Theory* **1** 96–101.

[15] Erdős, P. (1980) Some notes on Turán's mathematical work. *J. Approx. Theory* **29** no. 1, 2–5.

[16] Erdős, P. (1980) Some personal reminiscences of the mathematical work of Paul Turán. *Acta Arith.* **37** 4–8.

[17] Erdős, P. (1983) Preface. Personal reminiscences. *Studies in Pure Mathematics, To the memory of Paul Turán*, Birkhäuser, Basel-Boston, Mass., pp. 11–12.

[18] Erdős, P. (1989) On some aspects of my work with Gabriel Dirac. In *Graph theory in memory of G. A. Dirac*, Pap. Meet., Sandbjerg/Den. 1985. *Ann. Discrete Math* **41** 111–116.

[19] Erdős, P. (1981) On the combinatorial problems which I would most like to see solved. *Combinatorica* **1** 25-42.

[20] Erdős, P. (1982) Personal reminiscences and remarks on the mathematical work of Tibor Gallai. *Combinatorica* **2** 207–212.

[21] Erdős, P. (1985) E. Straus (1921–1983), Number theory (Winnipeg, Man., 1983). *Rocky Mountain J. Math.* **15** 331–341.

[22] Erdős, P. (1985) Ulam, the man and the mathematician. *J. Graph Theory* **9** no. 4, 445–449. Also appears in *Creation Math.* **19** (1986), 13–16.

[23] Erdős, P. (1987) My joint work with Richard Rado. In *Surveys in Combinatorics 1987* (New Cross, 1987), London Math. Soc. Lecture Note Ser. 123, pp. 53–80, Cambridge Univ. Press, Cambridge.

[24] Erdős, P. (1988) Recollections on Kurt Gödel. *Jahrb. Kurt-Gödel-Ges.* **1988**, 94–95.

[25] Erdős, P. (1989) Some personal and mathematical reminiscences of Kurt Mahler. *Aust. Math. Soc. Gaz.* **16** 1–2.

[26] Erdős, P. (1995) On the 120-th anniversary of the birth of Schur. *Geombinatorics* **5**, No.1, 4–5.

[27] Erdős, P. (1996) On some of my favorite theorems. In *Combinatorics: Paul Erdős is Eighty*, Vol. 2/2, Bolyai Society Math Studies, Keszthely, Hungary, pp. 97–132.

[28] Erdős, P. (1997) Some of my favorite problems and results. In *The Mathematics of Paul Erdős I* (R. L. Graham and J. Nešetřil, eds), Algorithms and Combinatorics, Vol. 13, Springer Verlag, pp. 47–67.

[29] Erdős, P. (ed.) (1989) *Collected Papers of Paul Turán*, Akadémiai Kiadó, Budapest. Vols 1–3.

[30] Erdős, P. and Kac, M. (1940) The Gaussian law of errors in the theory of additive number theoretic functions. *Amer. J. Math.* **62** 738–742.

[31] Erdős, P. and Kac, M. (1946) On certain limit theorems of the theory of probability. *Bull. Amer. Math. Soc* **52** 292–302.

[32] Erdős, P. and Rényi, A. (1963) Asymmetric graphs. *Acta Math. Hungar.* **14** 295–315.

[33] Erdős, P. and Rényi, A. (1961) On the strength of connectedness of a random graph. *Acta Math. Hungar.* **12** 261–267.

[34] Erdős, P. and Rényi, A. (1961) On the evolution of random graphs. *Bull. Inter. Stat. Inst. Tokyo* **38** 343–347.

[35] Erdős, P. and Rényi, A. (1960) On the evolution of random graphs. *MTA MKI Közl.* **5** 17–61.

[36] Erdős P., Rényi A. and Sós, V. T. (1966) On a problem of graph theory. *Studia Sci. Math. Hungar.* **1** 215–235. (Reprinted in [12].)

[37] Erdős, P. and Simonovits, M. (1969) Some extremal problems in graph theory. In *Combinatorial Theory and its Appl.*, Proc. Colloq. Math. Soc. János Bolyai **4** 377–384. (Reprinted in [12].)

[38] Erdős, P. and Spencer, J. (1974) *Probabilistic methods in combinatorics*, Academic Press, New York-London.

[39] Erdős, P. and Szekeres, Gy. (1935) A combinatorial problem in geometry. *Compositio Math.* **2** 463–470.

[40] Erdős, P. and Wintner, A. (1939) Additive functions and statistical independence. *Amer. J. Math.* **61** 713–721.

[41] Erdős, P. and Wintner, A. (1940) Additive functions and almost periodicity (B^2). *Amer. J. Math.* **62** 635–645.

[42] Hajnal, A. (1997) Paul Erdős' set theory. In *The Mathematics of Paul Erdős II* (R. L. Graham and J. Nešetřil, eds), Algorithms and Combinatorics, Vol. 14, Springer Verlag, pp. 352–393. (The part quoted here is on pp. 359–360.)

[43] Józsa, S. and Szemerédi, E. (1973) The number of unit distances on the plane. *Publ. Math. Soc. J. Bolyai* **10**, vol II, 939–950.

[44] Karoński, M. and Ruciński, A. (1997) The origins of the theory of random graphs. In *The Mathematics of Paul Erdős I* (R. L. Graham and J. Nešetřil, eds), Algorithms and Combinatorics, Vol. 13, Springer Verlag, pp. 311–336.

[45] Simonovits, M. (1997) Paul Erdős' influence on extremal graph theory. In *The Mathematics of Paul Erdős II* (R. L. Graham and J. Nešetřil, eds), Algorithms and Combinatorics, Vol. 14, Springer Verlag, pp. 148–192.

[46] Spencer, J., Szemerédi, E. and Trotter, W. T. (1984) Unit distances in the Euclidean plane. In *Graph Theory and Combinatorics* (B. Bollobás, ed.), Academic Press, New York, pp. 293–303.

[47] Sós, V. T. (1997) Paul Erdős, 1913-1996. *Aequationes Math.* 1997, 205–220.

[48] Szekeres, G. (1973) A Combinatorial Problem in Geometry. In the introductory part of [12], xix–xxii.

[49] Szemerédi, E. (1975) On a set containing no k elements in an arithmetic progression. *Acta Arithmetica* **27** 199–245.

[50] Turán, P. (1941) On an extremal problem in graph theory. *Matematikai Lapok* **48** 436–452 (in Hungarian). (See also [51], [29].)

[51] Turán, P. (1954) On the theory of graphs. *Colloq. Math.* **3** 19–30. (See also [29].)

[52] Turán, P. (1963) Paul Erdős is 50. *Matematikai Lapok* **14** 1–28 (in Hungarian). Reprinted in English in [29].

[53] Turán, P. (1977) A note of welcome. *Journal of Graph Theory* **1** 7–9.

Combinatorics, Probability and Computing (1999) 8, 1–6.
© 1999 Cambridge University Press

A Selection of Problems and Results
in Combinatorics

PAUL ERDŐS†

Mathematical Institute of the Hungarian Academy of Sciences,
13–15 Reáltanoda utca, 1053 Budapest, Hungary

In this note, I will present some new problems, and also some old ones which, in my opinion, have been undeservedly forgotten or neglected.

In this note, I present a number of problems in combinatorics that have attracted my attention recently. Some are new, while others are old but seem worthy of a fresh look.

1. This problem was started by Hajnal and myself in 1958. First, a few definitions from set theory. Let $\mathscr{S}$ be a set; to every finite subset $\mathscr{S}'$ of $\mathscr{S}$ we assign an element $f(\mathscr{S}') = x$ of $\mathscr{S}$ such that $x \notin \mathscr{S}'$. A subset $\mathscr{S}'$ of $\mathscr{S}$ is called *independent* or *free* if, for every subset $\mathscr{S}''$ of $\mathscr{S}'$, we have $f(\mathscr{S}'') \notin \mathscr{S}'$. In our first paper with Hajnal [8], we proved that if $|\mathscr{S}| < \aleph_\omega$ we can always find a mapping f from the set of finite subsets of $\mathscr{S}$ to $\mathscr{S}$ such that there is no infinite independent set. We could not decide whether there is always an infinite independent set if $|\mathscr{S}| = \aleph_\omega$. This problem is perhaps undecidable.

In a more recent paper [9], Hajnal and I investigated the following finite version of the problem. Let $|\mathscr{S}| = n < \aleph_0$ and let $h(n)$ be the largest integer such that, for every f, there is an independent subset $\mathscr{S}' \subset \mathscr{S}$, $|\mathscr{S}'| = h(n)$. Further, let $H(n)$ be the smallest integer for which there is a function f such that, for every $\mathscr{S}'' \subset \mathscr{S}$, $|\mathscr{S}''| \geqslant H(n)$, $F[\mathscr{S}''] = \mathscr{S}$, where $F[\mathscr{S}'']$ is the union of all the elements $f(\mathscr{S}''')$, with $\mathscr{S}''' \subset \mathscr{S}''$.

We proved

$$\frac{\log n}{\log 2} < H(n) < \frac{\log n}{\log 2} + \frac{(3 + o(1))\log\log n}{\log 2}.$$

We observed that

$$h(n) < \frac{\log n + 3\log\log n}{\log 2} + o(\log\log n),$$

but we only had a very poor lower bound for $h(n)$. Three years later Spencer and I [11] proved

$$h(n) > \frac{\log n - \log\log n}{\log 2} + o(\log\log n).$$

---

† This is the very last paper that Paul Erdős worked on before his death in Warsaw on 20 September 1996.

I think the problem was completely forgotten until a few days ago, when Gyárfás and I took up the problem again. We wanted to prove that

$$H(n) - \frac{\log n}{\log 2} \to \infty. \tag{1}$$

As far as I know (1) is still open; we could not even prove that, for $n = 2^k$,

$$H(n) \geqslant k + 1. \tag{2}$$

We first tried to prove that, for a set $\mathscr{S}$ of size 2^k, there is always a set $A \subset \mathscr{S}$ with $|A| = k$ such that

$$F[A] < 2^k - k. \tag{3}$$

It is easy to see that (3) implies (2), but we could not prove (3). In fact, we are sure that

$$F[A] < 2^k - k^c$$

for every c, if $n = 2^k$ is sufficiently large. It is not impossible that there is a set A of size k such that $F[A] < n(1 - \varepsilon)$ or even $F[A] = o(n)$.

2. Here is a recent problem that we formulated with Ralph Faudree. Let $G(2n)$ be a regular graph of $2n$ vertices and degree $n + 1$. Is it true that our $G(2n)$ has $c_1 2^{2n}$ subsets that are on a cycle? It is easy to see that there are at least $\left(\frac{1}{2} + o(1)\right) 2^{2n}$ sets that are not on a cycle. Also, if the regularity is replaced by the condition that the minimum degree is $n + 1$, the conjecture no longer holds.

3. An older problem which we posed before 1980 goes as follows. Every graph of $m \geqslant 2n + 1 \geqslant 7$ vertices and $\binom{2n+1}{2} - \binom{n}{2} - 1$ edges is the union of a bipartite graph and a graph of maximal degree at most $n - 1$. Faudree found a very nice proof if the number of vertices is exactly $2n + 1$, but the proof fails if the number of vertices is greater than $2n + 1$, and the problem is still open.

Let $\mathscr{C}_n$ denote the family of all cycles of lengths between 3 and $n + 3$ inclusive. In a paper that will soon appear, Faudree and I prove that, for $n \geqslant 2$,

$$r^*(\mathscr{C}_n, K_{1,n}) = \binom{2n + 1}{2} - \binom{n}{2}. \tag{4}$$

In human language, if $G(2n + 1, \binom{2n+1}{2} - \binom{n}{2})$ is a graph with $2n + 1$ vertices and $\binom{2n+1}{2} - \binom{n}{2}$ edges, then, if we colour the edges of our graph G red and blue, then either there is a red cycle C_m for every $3 \leqslant m \leqslant n + 3$ or there is a vertex incident with n blue edges.

Equation (4) is best possible. Our paper with Faudree [2] contains many further problems and results, but I have to leave these for the interested reader.

4. The next two problems arise from recent joint work of Gallai, Tuza and myself. Perhaps the most striking problem we have [5] is as follows. For G a graph, let $\alpha_1(G)$ denote the maximum cardinality of a set of edges that contains at most one edge from

every triangle of G (if G is triangle-free then $\alpha_1(G) = |E|$, the number of edges of G).
Similarly, let $\tau_1(G)$ be the minimum cardinality of a set of edges containing at least one
edge from every triangle of G. If G is triangle-free then $\tau_1(G) = 0$.

Is it true that, for every graph G on n vertices,

$$\alpha_1(G) + \tau_1(G) \leqslant \left[\frac{n^2}{4}\right]? \tag{5}$$

If true, (5) is probably quite difficult to prove. One difficulty seems to be that there are
several graphs of different structure for which there is equality in (5). Three examples are:
the complete graph $K(n)$, the complete bipartite graph $K(m, m)$, and the graph obtained
from $K(m, m)$ by adding one new vertex joined to every other.

In our paper [5], we prove several other interesting results and state some other
problems. We have tried to formulate related problems where the triangle is replaced by
a $K(r)$, $r > 3$, but so far we have not been quite successful.

5. In another triple paper with Gallai and Tuza [4], we investigate the following
problems. Define a *clique* of G to be a maximal complete subgraph of G. The *clique
transversal number* $\tau_C(G)$ is the smallest integer for which there is a set of $\tau_C(G)$ vertices
that intersects every clique of G. Our first problem was as follows. Let G run through all
graphs of n vertices: what is the maximum possible value of $\tau_C(G)$? Denote this maximum
by $f(n)$.

We conjecture that

$$f(n) = n - r(n), \tag{6}$$

where $r(n)$ is the largest integer such that every triangle-free graph contains an independent
set of $r(n)$ vertices. It is immediate that $f(n) \geqslant n - r(n)$; indeed, if we take a triangle-free
$G(n)$ with largest independent set of size $r(n)$, then we need to find a set intersecting
all the edges of our $G(n)$, which has size at least $n - r(n)$. The results of Ajtai, Komlós,
Szemerédi [1] and Kim [13] imply that

$$c_1(n \log n)^{1/2} < r(n) < c_2(n \log n)^{1/2}.$$

However, we could only prove

$$f(n) \leqslant n - (2n)^{1/2} + \frac{1}{2}. \tag{7}$$

As a first step we tried to prove

$$f(n) < n - s(n)n^{1/2},$$

for some function $s(n)$ tending to infinity. So far we have not been successful and in fact
we could not improve (7).

It seemed to us that if all cliques of G are large, then $\tau_C(G)$ will be smaller. For instance,
for $c > 0$, we would like estimates for the least $k_c(n)$ such that, if every clique of a graph
$G(n)$ has size $k_c(n)$, then $\tau_C(G(n)) < n(1 - c)$.

We proved that $k_c(n) > n^{c'(\log\log n)}$ is certainly needed to ensure $\tau_C(G(n)) < n(1-c)$. Perhaps $k(n) = n^{\alpha}$ will imply $\tau_C(G(n)) = o(n)$, but we did not prove anything about this. We did succeed in proving that, if every clique of $G(n)$ has size greater than k, then

$$\tau_C(G(n)) \leqslant n - (kn)^{1/2}.$$

Bollobás and I proved that if every clique has size at least $n+3-\lceil 2\sqrt{n}\rceil$ then $\tau_C(G(n)) = 1$, and that the value $n+3-\lceil 2\sqrt{n}\rceil$ is best possible.

Several further interesting problems are stated in our paper. Here I only state one of them. Let $G(n)$ be $K(4)$-free. How large a triangle-free induced subgraph must our $G(n)$ contain? An old result of Erdős and Szekeres [12] implies that our $G(n)$ contains an independent set of size $cn^{1/3}$, but perhaps it contains a much larger triangle-free induced subgraph. Several interesting related problems can be posed, but we must leave them to the interested reader. (I hope this set will not be empty.)

6. Now we discuss some problems of Gyárfás, Ruszinkó and myself [7]. Let $G(n)$ be a triangle-free graph of order n. We want to add as few edges to our $G(n)$ as possible, keeping it triangle-free, so as to make the diameter of the new graph equal to 2 (*i.e.*, so that no more edges can be added without creating a triangle). Denote by $h(G(n))$ the smallest number of edges having this property. We hoped that, if every vertex of $G(n)$ has degree $o(n^{1/2})$, then

$$h(G(n))/n^2 \to 0. \tag{8}$$

Simonovits showed that if the maximal degree is allowed to be $cn^{1/2}$ then (8) need not hold.

A few days ago we started to investigate the following related problem. Let $h_r(n)$ be the smallest integer such that, for every connected triangle-free graph $G(n)$, we can add a set of at most $h_r(n)$ edges to $G(n)$ and obtain a triangle-free graph of diameter at most r. We proved that

$$n - c\sqrt{n} \leqslant h_3(n) \leqslant n,$$

and later improved the lower bound to $n-c$. It is easy to see that $h_4(n) < n$, but we could not decide whether

$$h_4(n) < n(1-\varepsilon)$$

is true. To show that $h_5(n) < n(1-\varepsilon)$ is relatively easy, and we proved that $h_5(n) \leqslant \frac{n-1}{2}$.

7. Gyárfás and I have another paper [6] that will soon appear, and that I feel contains many interesting problems and results. Here is one of our favourite conjectures. If $r \geqslant 3$ and the edges of a $K(r^2 + 1)$ are coloured with r colours, then there exist $r + 1$ vertices with at least one colour not appearing on the edges they span. Clearly the conjecture fails for $r = 2$; we prove it for $r = 3$ and $r = 4$ – the proof is not quite trivial.

It is easy to show that, for infinitely many r, a $K(r^2)$ can be coloured with r colours so that every set of $r + 1$ vertices spans all the colours.

8.   In a recent paper [10], Hajnal, Tuza and I returned to a type of problem that Gallai and I considered many years ago.

Let X be a finite or infinite set, and $\mathcal{F}$ a system of subsets of X. We say that $\mathcal{F}$ is *r-uniform* if all elements E of $\mathcal{F}$ have $|E| = r$. We say that a subset T of X *represents* $\mathcal{F}$ if $T \cap E \neq \emptyset$ for all $E \in \mathcal{F}$. The *transversal number* (or *covering number*) of $\mathcal{F}$ is the minimum cardinality $\tau(\mathcal{F})$ of a set T which represents $\mathcal{F}$.

Let p, r, s, t be integers. We write

$$(p, s) \longrightarrow_r t$$

if, whenever $\mathcal{F}$ is an r-uniform set system such that every subsystem $\mathcal{F}' \subset \mathcal{F}$ on at most p elements has $\tau(\mathcal{F}') \leqslant s$, then we have $\tau(\mathcal{F}) \leqslant t$.

The first results on this topic were due to Gallai and myself [3]: for instance, we proved that

$$(2t + 2, t) \longrightarrow_2 t \text{ and } (2t + 1, t) \not\longrightarrow_2 t.$$

For $r \geqslant 3$, Hajnal, Tuza and I proved:

$$(3r - 3, 1) \longrightarrow_r \left\lceil \frac{r}{5} \right\rceil.$$

This is reasonably close to the truth, since we also showed that

$$(3r - 3, 1) \not\longrightarrow_r \left\lfloor \frac{3}{16}r + \frac{7}{8} \right\rfloor,$$

but we did not decide the best possible result here.

Several further problems and results are in our paper, but they are somewhat technical and we have to stop.

References

[1]   Ajtai, M., Komlós, J. and Szemerédi, E. (1980) A note on Ramsey numbers. *J. Combin. Theory Ser. A* **29** 354–360.

[2]   Erdős, P. and Faudree, R. Restricted size Ramsey numbers for cycles and stars. *Discrete Math.* To appear.

[3]   Erdős, P. and Gallai, T. (1961) On the maximal number of vertices representing the edges of a graph. *Magyar Tud. Akad. Mat. Kut. Inst. Közl.* **6** 181–203.

[4]   Erdős, P., Gallai, T. and Tuza, Zs. (1992) Covering the cliques of a graph with vertices. *Discrete Math.* **108** 279–289.

[5]   Erdős, P., Gallai, T. and Tuza, Zs. (1996) Covering and independence in triangle structures. *Discrete Math.* **150** 89–101.

[6]   Erdős, P. and Gyárfás, A. Split and balanced colorings of complete graphs. *Discrete Math.* To appear.

[7]   Erdős, P., Gyárfás, A. and Ruszinkó, M. How to decrease the diameter of triangle-free graphs? Submitted.

[8]   Erdős, P. and Hajnal, A. (1958) On the structure of set mappings. *Acta Math. Acad. Sci. Hungar.* **9** 111–133.

[9]   Erdős, P. and Hajnal, A. (1968) On a combinatorial problem. *Mat. Lapok* **19** 345–348. In Hungarian.

[10] Erdős, P., Hajnal, A. and Tuza, Zs. (1991) Local constraints ensuring small representing sets. *J. Combin. Theory Ser. A* **58** 78–84.

[11] Erdős, P. and Spencer, J. (1971) On a problem of Erdős and Hajnal. *Mat. Lapok* **22** 1–2. In Hungarian.

[12] Erdős, P. and Szekeres, G. (1935) A combinatorial problem in geometry. *Compositio Math.* **2** 463–470.

[13] Kim, J. H. (1995) The Ramsey number $R(3, t)$ has order of magnitude $t^2/\log t$. *Random Structures and Algorithms* **7** 173–207.

Combinatorics, Probability and Computing (1999) 8, 7–29.
© 1999 Cambridge University Press

Combinatorial Nullstellensatz

NOGA ALON†

Department of Mathematics, Raymond and Beverly Sackler Faculty of Exact Sciences,
Tel Aviv University, Tel Aviv, Israel
and
Institute for Advanced Study, Princeton, NJ 08540, USA
(e-mail: noga@math.tau.ac.il)

We present a general algebraic technique and discuss some of its numerous applications in combinatorial number theory, in graph theory and in combinatorics. These applications include results in additive number theory and in the study of graph colouring problems. Many of these are known results, to which we present unified proofs, and some results are new.

1. Introduction

Hilbert's Nullstellensatz (see, for instance, [60]) is the fundamental theorem that asserts that if F is an algebraically closed field, and $f, g_1, \ldots, g_m$ are polynomials in the ring of polynomials $F[x_1, \ldots, x_n]$, where f vanishes over all common zeros of $g_1, \ldots, g_m$, then there is an integer k and polynomials $h_1, \ldots, h_m$ in $F[x_1, \ldots, x_n]$ so that

$$f^k = \sum_{i=1}^{n} h_i g_i.$$

In the special case $m = n$, where each g_i is a univariate polynomial of the form $\prod_{s \in S_i}(x_i - s)$, a stronger conclusion holds, as follows.

Theorem 1.1. *Let F be an arbitrary field, and let $f = f(x_1, \ldots, x_n)$ be a polynomial in $F[x_1, \ldots, x_n]$. Let $S_1, \ldots, S_n$ be nonempty subsets of F and define $g_i(x_i) = \prod_{s \in S_i}(x_i - s)$. If f vanishes over all the common zeros of $g_1, \ldots, g_n$ (that is, if $f(s_1, \ldots, s_n) = 0$ for all $s_i \in S_i$), then there are polynomials $h_1, \ldots, h_n \in F[x_1, \ldots, x_n]$ satisfying $\deg(h_i) \leqslant \deg(f) - \deg(g_i)$ so*

† Research supported in part by a grant from the Israel Science Foundation, by a Sloan Foundation grant No. 96-6-2, by an NEC Research Institute grant, and by the Hermann Minkowski Minerva Center for Geometry at Tel Aviv University.

that

$$f = \sum_{i=1}^{n} h_i g_i.$$

Moreover, if $f, g_1, \ldots g_n$ lie in $R[x_1, \ldots, x_n]$ for some subring R of F then there are polynomials $h_i \in R[x_1, \ldots, x_n]$ as above.

As a consequence of the above one can prove the following.

Theorem 1.2. *Let F be an arbitrary field, and let $f = f(x_1, \ldots, x_n)$ be a polynomial in $F[x_1, \ldots, x_n]$. Suppose the degree $\deg(f)$ of f is $\sum_{i=1}^{n} t_i$, where each t_i is a nonnegative integer, and suppose the coefficient of $\prod_{i=1}^{n} x_i^{t_i}$ in f is nonzero. Then, if $S_1, \ldots, S_n$ are subsets of F with $|S_i| > t_i$, there are $s_1 \in S_1, s_2 \in S_2, \ldots, s_n \in S_n$ so that*

$$f(s_1, \ldots, s_n) \neq 0.$$

In this paper we prove these two theorems, which may be called *Combinatorial Nullstellensatz*, and describe several combinatorial applications of them. After presenting the (simple) proofs of the above theorems in Section 2, we show in Section 3 that the classical theorem of Chevalley and Warning on roots of systems of polynomials and the basic theorem of Cauchy and Davenport on the addition of residue classes follow as simple consequences. We proceed to describe additional applications in additive number theory and in graph theory and combinatorics in Sections 4, 5, 6, 7 and 8. Many of these applications are known results, proved here in a unified way, and some are new. There are several known results that assert that a combinatorial structure satisfies a certain combinatorial property if and only if an appropriate polynomial associated with it lies in a properly defined ideal. In Section 9 we apply our technique and obtain several new results of this form. Finally, Section 10 contains some concluding remarks and open problems.

2. The proofs of the two basic theorems

To prove Theorem 1.1 we need the following simple lemma proved, for example, in [13]. For the sake of completeness we include the short proof.

Lemma 2.1. *Let $P = P(x_1, x_2, \ldots, x_n)$ be a polynomial in n variables over an arbitrary field F. Suppose that the degree of P as a polynomial in x_i is at most t_i for $1 \leqslant i \leqslant n$, and let $S_i \subset F$ be a set of at least $t_i + 1$ distinct members of F. If $P(x_1, x_2, \ldots, x_n) = 0$ for all n-tuples $(x_1, \ldots, x_n) \in S_1 \times S_2 \times \cdots \times S_n$, then $P \equiv 0$.*

Proof.   We apply induction on n. For $n = 1$, the lemma is simply the assertion that a nonzero polynomial of degree t_1 in one variable can have at most t_1 distinct zeros. Assuming that the lemma holds for $n - 1$, we prove it for n ($n \geqslant 2$). Given a polynomial $P = P(x_1, \ldots, x_n)$ and sets S_i satisfying the hypotheses of the lemma, let us write P as a polynomial in x_n, that is,

$$P = \sum_{i=0}^{t_n} P_i(x_1,\ldots,x_{n-1})x_n^i,$$

where each P_i is a polynomial with x_j-degree bounded by t_j. For each fixed $(n-1)$-tuple

$$(x_1,\ldots,x_{n-1}) \in S_1 \times S_2 \times \cdots \times S_{n-1},$$

the polynomial in x_n obtained from P by substituting the values of $x_1,\ldots,x_{n-1}$ vanishes for all $x_n \in S_n$, and is thus identically 0. Thus $P_i(x_1,\ldots,x_{n-1}) = 0$ for all $(x_1,\ldots,x_{n-1}) \in S_1 \times \cdots \times S_{n-1}$. Hence, by the induction hypothesis, $P_i \equiv 0$ for all i, implying that $P \equiv 0$. This completes the induction and the proof of the lemma. $\quad\square$

Proof of Theorem 1.1. Define $t_i = |S_i| - 1$ for all i. By assumption,

$$f(x_1,\ldots,x_n) = 0 \quad \text{for every } n\text{-tuple} \quad (x_1,\ldots,x_n) \in S_1 \times S_2 \times \cdots \times S_n. \quad (2.1)$$

For each i, $1 \leqslant i \leqslant n$, let

$$g_i(x_i) = \prod_{s \in S_i}(x_i - s) = x_i^{t_i+1} - \sum_{j=0}^{t_i} g_{ij}x_i^j.$$

Observe that,

$$\text{if } x_i \in S_i, \text{ then } g_i(x_i) = 0; \text{ that is, } x_i^{t_i+1} = \sum_{j=0}^{t_i} g_{ij}x_i^j. \quad (2.2)$$

Let $\bar{f}$ be the polynomial obtained by writing f as a linear combination of monomials and replacing, repeatedly, each occurrence of $x_i^{f_i}$ $(1 \leqslant i \leqslant n)$, where $f_i > t_i$, by a linear combination of smaller powers of x_i, using the relations (2.2). The resulting polynomial $\bar{f}$ is clearly of degree at most t_i in x_i, for each $1 \leqslant i \leqslant n$, and is obtained from f by subtracting from it products of the form $h_i g_i$, where the degree of each polynomial $h_i \in F[x_1,\ldots,x_n]$ does not exceed $\deg(f) - \deg(g_i)$ (and where the coefficients of each h_i are in the smallest ring containing all coefficients of f and $g_1,\ldots,g_n$). Moreover, $\bar{f}(x_1,\ldots,x_n) = f(x_1,\ldots,x_n)$, for all $(x_1,\ldots,x_n) \in S_1 \times \cdots \times S_n$, since the relations (2.2) hold for these values of $x_1,\ldots,x_n$. Therefore, by (2.1), $\bar{f}(x_1,\ldots,x_n) = 0$ for every n-tuple $(x_1,\ldots,x_n) \in S_1 \times \cdots \times S_n$ and hence, by Lemma 2.1, $\bar{f} \equiv 0$. This implies that $f = \sum_{i=1}^{n} h_i g_i$, and completes the proof. $\quad\square$

Proof of Theorem 1.2. Clearly we may assume that $|S_i| = t_i + 1$ for all i. Suppose the result is false, and define $g_i(x_i) = \prod_{s \in S_i}(x_i - s)$. By Theorem 1.1 there are polynomials $h_1,\ldots,h_n \in F[x_1,\ldots,x_n]$ satisfying $\deg(h_j) \leqslant \sum_{i=1}^{n} t_i - \deg(g_j)$ so that

$$f = \sum_{i=1}^{n} h_i g_i.$$

By assumption, the coefficient of $\prod_{i=1}^{n} x_i^{t_i}$ in the left-hand side is nonzero, and hence so is the coefficient of this monomial in the right-hand side. However, the degree of $h_i g_i = h_i \prod_{s \in S_i}(x_i - s)$ is at most $\deg(f)$, and if there are any monomials of degree $\deg(f)$ in it they are divisible by $x_i^{t_i+1}$. It follows that the coefficient of $\prod_{i=1}^{n} x_i^{t_i}$ in the right-hand side is zero, and this contradiction completes the proof. $\quad\square$

3. Two classical applications

The following theorem, conjectured by Artin in 1934, was proved by Chevalley in 1935 and extended by Warning in 1935. Here we present a very short proof using our Theorem 1.2 above. For simplicity, we restrict ourselves to the case of finite prime fields, though the proof easily extends to arbitrary finite fields.

Theorem 3.1 (e.g., [53]). *Let p be a prime, and let*

$$P_1 = P_1(x_1,\ldots,x_n),\ P_2 = P_2(x_1,\ldots,x_n),\ldots,\ P_m = P_m(x_1,\ldots,x_n)$$

be m polynomials in the ring $Z_p[x_1,\ldots,x_n]$. If $n > \sum_{i=1}^m \deg(P_i)$ and the polynomials P_i have a common zero $(c_1,\ldots,c_n)$, then they have another common zero.

Proof. Suppose this is false, and define

$$f = f(x_1,\ldots,x_n) = \prod_{i=1}^m (1 - P_i(x_1,\ldots,x_n)^{p-1}) - \delta \prod_{j=1}^n \prod_{c \in Z_p, c \neq c_j} (x_j - c),$$

where δ is chosen so that

$$f(c_1,\ldots,c_n) = 0. \tag{3.1}$$

Note that this determines the value of δ, and this value is nonzero. Note also that

$$f(s_1,\ldots,s_n) = 0 \tag{3.2}$$

for all $s_i \in Z_p$. Indeed, this is certainly true, by (3.1), if $(s_1,\ldots,s_n) = (c_1,\ldots,c_n)$. For other values of $(s_1,\ldots,s_n)$, there is, by assumption, a polynomial P_j that does not vanish on $(s_1,\ldots,s_n)$, implying that $1 - P_j(s_1,\ldots,s_n)^{p-1} = 0$. Similarly, since $s_i \neq c_i$ for some i, the product $\prod_{c \in Z_p, c \neq c_i} (s_i - c)$ is zero and hence so is the value of $f(s_1,\ldots,s_n)$.

Define $t_i = p - 1$ for all i and note that the coefficient of $\prod_{i=1}^n x_i^{t_i}$ in f is $-\delta \neq 0$, since the total degree of

$$\prod_{i=1}^m (1 - P_i(x_1,\ldots,x_n)^{p-1})$$

is $(p-1)\sum_{i=1}^m \deg(P_i) < (p-1)n$. Therefore, by Theorem 1.2 with $S_i = Z_p$ for all i, we conclude that there are $s_1,\ldots,s_n \in Z_p$ for which $f(s_1,\ldots,s_n) \neq 0$, contradicting (3.2) and completing the proof. □

The Cauchy–Davenport theorem, which has numerous applications in additive number theory, is the following.

Theorem 3.2 ([21]). *If p is a prime, and A, B are two nonempty subsets of Z_p, then*

$$|A + B| \geqslant \min\{p, |A| + |B| - 1\}.$$

Cauchy proved this theorem in 1813, and applied it to give a new proof to a lemma of Lagrange in his well-known 1770 paper that shows that any integer is a sum of four squares. Davenport formulated the theorem as a discrete analogue of a conjecture of Khintchine (which was proved a few years later by H. Mann) about the Schnirelman density of the sum of two sequences of integers. There are numerous extensions of this result: see, for instance, [46]. The proofs of Theorem 3.2 given by Cauchy and Davenport are based on the same combinatorial idea, and apply induction on $|B|$. A different, algebraic, proof has recently been found by the authors of [10] and [11], and its main advantage is that it extends easily and gives several related results. As shown below, this proof can be described as a simple application of Theorem 1.2.

Proof of Theorem 3.2. If $|A|+|B| > p$ the result is trivial, since in this case for every $g \in Z_p$ the two sets A and $g - B$ intersect, implying that $A + B = Z_p$. Assume, therefore, that $|A| + |B| \leqslant p$ and suppose the result is false and $|A + B| \leqslant |A| + |B| - 2$. Let C be a subset of Z_p satisfying $A + B \subset C$ and $|C| = |A| + |B| - 2$. Define $f = f(x, y) = \prod_{c \in C}(x + y - c)$ and observe that, by the definition of C,

$$f(a, b) = 0 \text{ for all } a \in A, b \in B. \tag{3.3}$$

Put $t_1 = |A| - 1, t_2 = |B| - 1$ and note that the coefficient of $x^{t_1} y^{t_2}$ in f is the binomial coefficient $\binom{|A|+|B|-2}{|A|-1}$ which is nonzero in Z_p, since $|A| + |B| - 2 < p$. Therefore, by Theorem 1.2 (with $n = 2, S_1 = A, S_2 = B$), there is an $a \in A$ and a $b \in B$ so that $f(a, b) \neq 0$, contradicting (3.3) and completing the proof. $\quad\square$

4. Restricted sums

The first theorem in this section is a general result, first proved in [11]. Here we observe that it is a simple consequence of Theorem 1.2 above. We also describe some of its applications, proved in [11], which are extensions of the Cauchy–Davenport theorem.

Let p be a prime. For a polynomial $h = h(x_0, x_1, \ldots, x_k)$ over Z_p and for subsets $A_0, A_1, \ldots, A_k$ of Z_p, define

$$\oplus_h \sum_{i=0}^{k} A_i = \{a_0 + a_1 + \cdots + a_k : a_i \in A_i, \ h(a_0, a_1, \ldots, a_k) \neq 0\}.$$

Theorem 4.1 ([11]). *Let p be a prime and let $h = h(x_0, \ldots, x_k)$ be a polynomial over Z_p. Let $A_0, A_1, \ldots, A_k$ be nonempty subsets of Z_p, where $|A_i| = c_i + 1$, and define $m = \sum_{i=0}^{k} c_i - \deg(h)$. If the coefficient of $\prod_{i=0}^{k} x_i^{c_i}$ in*

$$(x_0 + x_1 + \cdots + x_k)^m h(x_0, x_1, \ldots, x_k)$$

is nonzero (in Z_p), then

$$\left| \oplus_h \sum_{i=0}^{k} A_i \right| \geqslant m + 1$$

(and hence $m < p$).

Proof. Suppose the assertion is false, and let E be a (multi-) set of m (not necessarily distinct) elements of Z_p that contains the set $\oplus_h \sum_{i=0}^k A_i$. Let $Q = Q(x_0, \ldots, x_k)$ be the polynomial defined as follows:

$$Q(x_0, \ldots, x_k) = h(x_0, x_1, \ldots x_k) \prod_{e \in E} (x_0 + \cdots + x_k - e).$$

Note that

$$Q(x_0, \ldots, x_k) = 0 \quad \text{for all} \quad (x_0, \ldots, x_k) \in (A_0, \ldots, A_k). \tag{4.1}$$

This is because, for each such $(x_0, \ldots, x_k)$, either $h(x_0, \ldots, x_k) = 0$ or $x_0 + \cdots + x_k \in \oplus_h \sum_{i=0}^k A_i \subset E$. Note also that $\deg(Q) = m + \deg(h) = \sum_{i=0}^k c_i$ and hence the coefficient of the monomial $x_0^{c_0} \cdots x_k^{c_k}$ in Q is the same as that of this monomial in the polynomial $(x_0 + \cdots + x_k)^m h(x_0, \ldots, x_k)$, which is nonzero, by assumption.

By Theorem 1.2 there are $x_0 \in A_0$, $x_1 \in A_1, \ldots, x_k \in A_k$ such that $Q(x_0, x_1, \ldots, x_k) \neq 0$, contradicting (4.1) and completing the proof. $\qquad\square$

One of the applications of the last theorem is the following.

Proposition 4.2. *Let p be a prime, and let $A_0, A_1, \ldots, A_k$ be nonempty subsets of the cyclic group Z_p. If $|A_i| \neq |A_j|$ for all $0 \leqslant i < j \leqslant k$ and $\sum_{i=0}^k |A_i| \leqslant p + \binom{k+2}{2} - 1$, then*

$$\left| \{a_0 + a_1 + \cdots + a_k : a_i \in A_i, a_i \neq a_j \text{ for all } i \neq j\} \right| \geqslant \sum_{i=0}^k |A_i| - \binom{k+2}{2} + 1.$$

Note that the very special case of this proposition in which $k = 1$, $A_0 = A$ and $A_1 = A - \{a\}$ for an arbitrary element $a \in A$ implies that, if $A \subset Z_p$ and $2|A| - 1 \leqslant p + 2$, then the number of sums $a_1 + a_2$ with $a_1, a_2 \in A$ and $a_1 \neq a_2$ is at least $2|A| - 3$. This easily implies the following theorem, conjectured by Erdős and Heilbronn in 1964 (see, for instance, [26]). Special cases of this conjecture have been proved by various researchers [50, 44, 51, 30] and the full conjecture has recently been proved by Dias da Silva and Hamidoune [22], using some tools from linear algebra and the representation theory of the symmetric group.

Theorem 4.3 ([22]). *If p is a prime, and A is a nonempty subset of Z_p, then*

$$\left| \{a + a' : a, a' \in A, a \neq a'\} \right| \geqslant \min\{p, 2|A| - 3\}.$$

In order to deduce Proposition 4.2 from Theorem 4.1 we need the following lemma, which can be easily deduced from the known results about the Ballot problem (see, for instance, [45]), as well as from the known connection between this problem and the hook formula for the number of Young tableaux of a given shape. A simple, direct proof is given in [11].

Lemma 4.4. *Let $c_0, \ldots, c_k$ be nonnegative integers and suppose that $\sum_{i=0}^{k} c_i = m + \binom{k+1}{2}$, where m is a nonnegative integer. Then the coefficient of $\prod_{i=0}^{k} x_i^{c_i}$ in the polynomial*

$$(x_0 + x_1 + \cdots + x_k)^m \prod_{k \geqslant i > j \geqslant 0} (x_i - x_j)$$

is

$$\frac{m!}{c_0! c_1! \ldots c_k!} \prod_{k \geqslant i > j \geqslant 0} (c_i - c_j).$$

$\square$

Let p be a prime, and let $A_0, A_1, \ldots, A_k$ be nonempty subsets of the cyclic group Z_p. Define

$$\oplus_{i=0}^{k} A_i = \{a_0 + a_1 + \cdots + a_k : a_i \in A_i, a_i \neq a_j \text{ for all } i \neq j\}.$$

In this notation, the assertion of Proposition 4.2 is that if $|A_i| \neq |A_j|$ for all $0 \leqslant i < j \leqslant k$ and $\sum_{i=0}^{k} |A_i| \leqslant p + \binom{k+2}{2} - 1$ then

$$|\oplus_{i=0}^{k} A_i| \geqslant \sum_{i=0}^{k} |A_i| - \binom{k+2}{2} + 1.$$

Proof of Proposition 4.2. Define

$$h(x_0, \ldots, x_k) = \prod_{k \geqslant i > j \geqslant 0} (x_i - x_j),$$

and note that, for this h, the sum $\oplus_{i=0}^{k} A_i$ is precisely the sum $\oplus_h \sum_{i=0}^{k} A_i$. Suppose $|A_i| = c_i + 1$ and put

$$m = \sum_{i=0}^{k} c_i - \binom{k+1}{2} \quad \left(= \sum_{i=0}^{k} |A_i| - \binom{k+2}{2} \right).$$

By assumption $m < p$ and by Lemma 4.4, the coefficient of $\prod_{i=0}^{k} x_i^{c_i}$ in

$$h \cdot (x_0 + \cdots + x_k)^m$$

is

$$\frac{m!}{c_0! c_1! \ldots c_k!} \prod_{k \geqslant i > j \geqslant 0} (c_i - c_j),$$

which is nonzero modulo p, since $m < p$ and the numbers c_i are pairwise distinct. Since $m = \sum_{i=0}^{k} c_i + \deg(h)$, the desired result follows from Theorem 4.1. $\square$

An easy consequence of Proposition 4.2 is the following. See [11] for the detailed proof.

Theorem 4.5. *Let p be a prime, and let $A_0, \ldots, A_k$ be nonempty subsets of Z_p, where $|A_i| = b_i$, and suppose $b_0 \geqslant b_1 \ldots \geqslant b_k$. Define $b_0', \ldots, b_k'$ by*

$$b_0' = b_0 \quad \text{and} \quad b_i' = \min\{b_{i-1}' - 1, b_i\}, \text{ for } 1 \leqslant i \leqslant k. \tag{4.2}$$

If $b'_k > 0$ then

$$| \oplus_{i=0}^k A_i | \geqslant \min\left\{ p, \sum_{i=0}^k b'_i - \binom{k+2}{2} + 1 \right\}.$$

Moreover, the above estimate is sharp for all possible values of $p \geqslant b_0 \geqslant \cdots \geqslant b_k$.

The following result of Dias da Silva and Hamidoune [22] is a simple consequence of (a special case of) the above theorem.

Theorem 4.6 ([22]). *Let p be a prime and let A be a nonempty subset of Z_p. Let $s^\wedge A$ denote the set of all sums of s distinct elements of A. Then $|s^\wedge A| \geqslant \min\{p, s|A| - s^2 + 1\}$.*

Proof. If $|A| < s$ there is nothing to prove. Otherwise put $s = k+1$ and apply Theorem 4.5 with $A_i = A$ for all i. Here $b'_i = |A| - i$ for all $0 \leqslant i \leqslant k$ and hence

$$|(k+1)^\wedge A| = | \oplus_{i=0}^k A_i | \quad \geqslant \quad \min\left\{ p, \sum_{i=0}^k (|A| - i) - \binom{k+2}{2} + 1 \right\}$$

$$= \quad \min\left\{ p, (k+1)|A| - \binom{k+1}{2} - \binom{k+2}{2} + 1 \right\}$$

$$= \quad \min\{ p, (k+1)|A| - (k+1)^2 + 1 \}. \qquad \square$$

Another easy application of Theorem 4.1 is the following result, proved in [10].

Proposition 4.7. *If p is a prime and A, B are two nonempty subsets of Z_p, then*

$$|\{a + b : a \in A, b \in B, ab \neq 1\}| \geqslant \min\{p, |A| + |B| - 3\}.$$

The proof is by applying Theorem 4.1 with $k = 1$, $h = x_0 x_1 - 1$, $A_0 = A$, $A_1 = B$, and $m = |A| + |B| - 4$. It is also shown in [10] that the above estimate is tight in all nontrivial cases. Additional extensions of the above proposition appear in [11].

5. Set addition in vector spaces over prime fields

A triple (r, s, n) of positive integers satisfies the *Hopf–Stiefel condition* if

$$\binom{n}{k}$$ *is even for every integer k satisfying $n - r < k < s$.*

This condition arises in topology. However, studying the combinatorial aspects of the well-known Hurwitz problem, Yuzvinsky [61] showed that it has an interesting relation to a natural additive problem. He proved that in a vector space of infinite dimension over $GF(2)$, there exist two subsets $A, B \subset V$ satisfying $|A| = r$, $|B| = s$ and $|A + B| \leqslant n$ if and only if the triple (r, s, n) satisfies the Hopf–Stiefel condition.

Eliahou and Kervaire [24] have shown very recently that this can be proved using the algebraic technique of [10] and [11], and generalized this result to an arbitrary prime p, thus obtaining a common generalization of Yuzvinsky's result and the Cauchy–Davenport theorem. Here is a description of their result, and a quick derivation of it from Theorem 1.2. It is worth noting that the same result also follows from the main result of Bollobás and Leader in [19], proved by a different, more combinatorial, approach.

Let us say that a triple (r, s, n) of positive integers satisfies the *Hopf–Stiefel condition with respect to a prime p* if

$$\binom{n}{k} \text{ is divisible by } p \text{ for every integer } k \text{ satisfying } n - r < k < s. \tag{5.1}$$

Let $\beta_p(r, s)$ denote the smallest integer n for which the triple (r, s, n) satisfies (5.1). We note that it is not difficult to give a recursive formula for $\beta_p(r, s)$, which enables one to compute it quickly, given the representation of r and s in basis p.

Theorem 5.1 ([24]). *If A and B are two finite nonempty subsets of a vector space V over $GF(p)$, and $|A| = r$, $|B| = s$, then $|A + B| \geqslant \beta_p(r, s)$.*

Proof. We may assume that V is finite, and identify it with the finite field F_q of the same cardinality over $GF(p)$. Viewing A and B as subsets of F_q, define $C = A + B$, and assume the assertion is false and $|C| = n < \beta_p(r, s)$. As in the previous section, define

$$Q(x, y) = \prod_{c \in C} (x + y - c),$$

where Q is a polynomial over F_q, and observe that $Q(a, b) = 0$ for all $a \in A, b \in B$. By the definition of $\beta_p(r, s)$, there is some k satisfying $n - r < k < s$ such that $\binom{n}{k}$ is not divisible by p. Therefore, the coefficient of $x^{n-k} y^k$ in the above polynomial is not zero, and since $|A| = r > n - k$, $|B| = s > k$, there are, by Theorem 1.2, $a \in A$ and $b \in B$ such that $Q(a, b) \neq 0$: a contradiction. This completes the proof. $\square$

The authors of [24] have also shown that the estimate in Theorem 5.1 is sharp for all possible r and s. In fact, if A is the set of r vectors whose coordinates correspond to the p-adic representation of the integers $0, 1, \ldots, r - 1$, and B is the set of s vectors whose coordinates correspond to the p-adic representation of the integers $0, 1, \ldots, s - 1$, it is not too difficult to check that $A + B$ is the set of of all vectors whose coordinates correspond to the p-adic representation of the integers $0, 1, \ldots, \beta_p(r, s) - 1$. For more details and several extensions, see [24].

6. Graphs, subgraphs and cubes

A well-known conjecture of Berge and Sauer, proved by Taškinov [54], asserts that any simple 4-regular graph contains a 3-regular subgraph. This assertion is easily seen to be false for graphs with multiple edges, but, as shown in [6], one extra edge suffices to ensure a 3-regular subgraph in this more general case as well. This follows from the case $p = 3$ in the following result, which, as shown below, can be derived quickly from Theorem 1.2.

Theorem 6.1 ([6]). *For any prime p, any loopless graph $G = (V, E)$ with average degree bigger than $2p - 2$ and maximum degree at most $2p - 1$ contains a p-regular subgraph.*

Proof. Let $(a_{v,e})_{v\in V, e\in E}$ denote the incidence matrix of G defined by $a_{v,e} = 1$ if $v \in e$ and $a_{v,e} = 0$ otherwise. Associate each edge e of G with a variable x_e and consider the polynomial

$$F = \prod_{v\in V}\left[1 - \left(\sum_{e\in E} a_{v,e}x_e\right)^{p-1}\right] - \prod_{e\in E}(1 - x_e),$$

over $GF(p)$. Notice that the degree of F is $|E|$, since the degree of the first product is at most $(p-1)|V| < |E|$, by the assumption on the average degree of G. Moreover, the coefficient of $\prod_{e\in E} x_e$ in F is $(-1)^{|E|+1} \neq 0$. Therefore, by Theorem 1.2, there are values $x_e \in \{0,1\}$ such that $F(x_e : e \in E) \neq 0$. By the definition of F, the above vector $(x_e : e \in E)$ is not the zero vector, since for this vector $F = 0$. In addition, for this vector, $\sum_{e\in E} a_{v,e}x_e$ is zero modulo p for every v, since otherwise F would vanish at this point. Therefore, in the subgraph consisting of all edges $e \in E$ for which $x_e = 1$ all degrees are divisible by p, and since the maximum degree is smaller than $2p$ all positive degrees are precisely p, as needed. $\qquad\square$

The assertion of Theorem 6.1 is proved in [6] for prime powers p as well, but it is not known if it holds for every integer p. Combining this result with some additional combinatorial arguments, one can show that for every $k \geqslant 4r$, every loopless k-regular graph contains an r-regular subgraph. For more details and additional results, see [6].

Erdős and Sauer (see, for instance, [17], page 399) raised the problem of estimating the maximum number of edges in a simple graph on n vertices that contains no 3-regular subgraph. They conjectured that for every positive ϵ this number does not exceed $n^{1+\epsilon}$, provided n is sufficiently large as a function of ϵ. This has been proved by Pyber [48], using Theorem 6.1. He proved that any simple graph on n vertices with at least $200n \log n$ edges contains a subgraph with maximum degree 5 and average degree more than 4. This subgraph contains, by Theorem 6.1, a 3-regular subgraph. On the other hand, Pyber, Rödl and Szemerédi [49] proved, by probabilistic arguments, that there are simple graphs on n vertices with at least $\Omega(n \log \log n)$ edges that contain no 3-regular subgraphs. Thus Pyber's estimate is not far from being best possible.

Here is another application of Theorem 1.2, which is not very natural, but demonstrates its versatility.

Proposition 6.2. *Let p be a prime, and let $G = (V, E)$ be a graph on a set of $|V| > d(p-1)$ vertices. Then there is a nonempty subset U of vertices of G such that the number of cliques of d vertices of G that intersect U is 0 modulo p.*

Proof. For each subset I of vertices of G, let $K(I)$ denote the number of copies of K_d in G that contain I. Associate each vertex $v \in V$ with a variable x_v, and consider the polynomial

$$F = \prod_{v\in V}(1 - x_v) - 1 + G,$$

where

$$G = \left[\sum_{\emptyset \neq I \subset V} (-1)^{|I|+1} K(I) \prod_{i \in I} x_i \right]^{p-1}$$

over $GF(p)$. Since $K(I)$ is obviously zero for all I of cardinality bigger than d, the degree of this polynomial is $|V|$, as the degree of G is at most $d(p-1) < |V|$. Moreover, the coefficient of $\prod_{v \in V} x_v$ in F is $(-1)^{|V|} \neq 0$. Therefore, by Theorem 1.2, there are $x_v \in \{0, 1\}$ for which $F(x_v : v \in V) \neq 0$. Since F vanishes on the all zero vector, it follows that not all numbers x_v are zero, and hence that $G(x_v : v \in V) \neq 1$, implying, by Fermat's Little Theorem, that

$$\sum_{\emptyset \neq I \subset V} (-1)^{|I|+1} K(I) \prod_{i \in I} x_i \equiv 0 (\bmod\ p).$$

However, the left-hand side of the last congruence is precisely the number of copies of K_d that intersect the set $U = \{v : x_v = 1\}$, by the Inclusion-Exclusion formula. Since U is nonempty, the desired result follows. $\qquad\square$

The assertion of the last proposition can be proved for prime powers p as well. See also [8] and [4] for some related results. Some versions of these results arise in the study of the minimum possible degree of a polynomial that represents the OR function of n variables in the sense discussed in [56] and its references.

We close this section with a simple geometric result, proved in [7], answering a question of Komját. As shown below, this result is also a simple consequence of Theorem 1.2.

Theorem 6.3 ([7]). *Let $H_1, H_2, \ldots, H_m$ be a family of hyperplanes in R^n that cover all vertices of the unit cube $\{0, 1\}^n$ but one. Then $m \geq n$.*

Proof. Clearly we may assume that the uncovered vertex is the all zero vector. Let $(a_i, x) = b_i$ be the equation defining H_i, where $x = (x_1, x_2, \ldots, x_n)$, and (a, b) is the inner product between the two vectors a and b. Note that for every i, $b_i \neq 0$, since H_i does not cover the origin. Assume the assertion is false and $m < n$, and consider the polynomial

$$P(x) = (-1)^{n+m+1} \prod_{j=1}^{m} b_j \prod_{i=1}^{n} (x_i - 1) - \prod_{i=1}^{m} [(a_i, x) - b_i].$$

The degree of this polynomial is clearly n, and the coefficient of $\prod_{i=1}^{n} x_i$ in it is

$$(-1)^{n+m+1} \prod_{j=1}^{m} b_j \neq 0.$$

Therefore, by Theorem 1.2 there is a point $x \in \{0, 1\}^n$ for which $P(x) \neq 0$. This point is not the all zero vector, as P vanishes on it, and therefore it is some other vertex of the cube. But in this case $(a_i, x) - b_i = 0$ for some i (as the vertex is covered by some H_i), implying that P does vanish on this point: a contradiction. $\qquad\square$

The above result is clearly tight. Several extensions are proved in [7].

7. Graph colouring

Graph colouring is arguably the most popular subject in graph theory. An interesting variant of the classical problem of colouring properly the vertices of a graph with the minimum possible number of colours arises when one imposes some restrictions on the colours available for every vertex. This variant received a considerable amount of attention that led to several fascinating conjectures and results, and its study combines interesting combinatorial techniques with powerful algebraic and probabilistic ideas. The subject, initiated independently by Vizing [59] and by Erdős, Rubin and Taylor [28], is usually known as the study of the *choosability* properties of a graph. Tarsi and the author developed in [13] an algebraic technique that has already been applied by various researchers to solve several problems in this area as well as problems dealing with traditional graph colouring. In this section we observe that the basic results of this technique can be derived from Theorem 1.2, and describe various applications. More details on some of these applications can be found in the survey [2].

We start with some notation and background. A *vertex colouring* of a graph G is an assignment of a colour to each vertex of G. The colouring is *proper* if adjacent vertices receive distinct colours. The *chromatic number* $\chi(G)$ of G is the minimum number of colours used in a proper vertex colouring of G. An *edge colouring* of G is, similarly, an assignment of a colour to each edge of G. It is *proper* if adjacent edges receive distinct colours. The minimum number of colours in a proper edge colouring of G is the *chromatic index* $\chi'(G)$ of G. This is clearly equal to the chromatic number of the line graph of G.

If $G = (V, E)$ is a (finite, directed or undirected) graph, and f is a function that assigns to each vertex v of G a positive integer $f(v)$, we say that G is f-*choosable* if, for every assignment of sets of integers $S(v) \subset Z$ to all the vertices $v \in V$, where $|S(v)| = f(v)$ for all v, there is a proper vertex colouring $c : V \mapsto Z$ so that $c(v) \in S(v)$ for all $v \in V$. The graph G is k-*choosable* if it is f-choosable for the constant function $f(v) \equiv k$. The *choice number* of G, denoted $ch(G)$, is the minimum integer k such that G is k-choosable. Obviously, this number is at least the classical chromatic number $\chi(G)$ of G. The choice number of the line graph of G, which we denote here by $ch'(G)$, is usually called the *list chromatic index* of G, and it is clearly at least the chromatic index $\chi'(G)$ of G.

As observed by various researchers, there are many graphs G for which the choice number $ch(G)$ is strictly larger than the chromatic number $\chi(G)$. A simple example demonstrating this fact is the complete bipartite graph $K_{3,3}$. If $\{u_1, u_2, u_3\}$ and $\{v_1, v_2, v_3\}$ are its two vertex-classes and $S(u_i) = S(v_i) = \{1, 2, 3\} \setminus \{i\}$, then there is no proper vertex colouring assigning to each vertex w a colour from its class $S(w)$. Therefore, the choice number of this graph exceeds its chromatic number. In fact, it is not difficult to show that, for any $k \geqslant 2$, there are bipartite graphs whose choice number exceeds k. Moreover, in [2] it is proved, using probabilistic arguments, that for every k there is some finite $c(k)$ so that the choice number of every simple graph with minimum degree at least $c(k)$ exceeds k.

In view of this, the following conjecture, suggested independently by various researchers including Vizing, Albertson, Collins, Tucker and Gupta, but apparently first published by Bollobás and Harris ([18]), is somewhat surprising.

Conjecture 7.1 (The List Colouring Conjecture). *For every graph G,* ch$'(G) = \chi'(G)$.

This conjecture asserts that for *line graphs* there is no gap at all between the choice number and the chromatic number. Many of the most interesting results in the area are proofs of special cases of this conjecture, which is still wide open. An asymptotic version of it, however, has been proven by Kahn [39] using probabilistic arguments: for simple graphs of maximum degree d, ch$'(G) = (1 + o(1))d$, where the $o(1)$-term tends to zero as d tends to infinity. Since in this case $\chi'(G)$ is either d or $d + 1$, by Vizing's theorem [58], this shows that the List Colouring Conjecture is asymptotically nearly correct.

The *graph polynomial* $f_G = f_G(x_1, x_2, \ldots, x_n)$ of a directed or undirected graph $G = (V, E)$ on a set $V = \{v_1, \ldots, v_n\}$ of n vertices is defined by $f_G(x_1, x_2, \ldots, x_n) = \Pi\{(x_i - x_j) : i < j, \{v_i, v_j\} \in E\}$. This polynomial has been studied by various researchers, starting with Petersen [47] in 1891. See also, for example, [52] and [41].

A subdigraph H of a directed graph D is called *Eulerian* if the in-degree $d_H^-(v)$ of every vertex v of H is equal to its out-degree $d_H^+(v)$. Note that we do not assume that H is connected. H is *even* if it has an even number of edges; otherwise, it is *odd*. Let $EE(D)$ and $EO(D)$ denote the numbers of even and odd Eulerian subgraphs of D, respectively. (For convenience we agree that the empty subgraph is an even Eulerian subgraph.) The following result is proved in [13].

Theorem 7.2. *Let $D = (V, E)$ be an orientation of an undirected graph G, denote $V = \{1, 2, \ldots, n\}$ and define $f : V \mapsto Z$ by $f(i) = d_i + 1$, where d_i is the out-degree of i in D. If $EE(D) \neq EO(D)$, then D is f-choosable.*

Proof (sketch). For $1 \leqslant i \leqslant n$, let $S_i \subset Z$ be a set of $d_i + 1$ distinct integers. The existence of a proper colouring of D assigning to each vertex i a colour from its list S_i is equivalent to the existence of colours $c_i \in S_i$ such that $f_G(c_1, c_2, \ldots, c_n) \neq 0$.

Since the degree of f_G is $\sum_{i=1}^n d_i$, it suffices to show that the coefficient of $\prod_{i=1}^n x_i^{d_i}$ in f_G is nonzero in order to deduce the existence of such colours c_i from Theorem 1.2. This can be done by interpreting this coefficient combinatorially.

It is not too difficult to see that the coefficients of the monomials that appear in the standard representation of f_G as a linear combination of monomials can be expressed in terms of the orientations of G as follows. Call an orientation D of G *even* if the number of its directed edges (i, j) with $i > j$ is even; otherwise call it *odd*. For nonnegative integers $d_1, d_2, \ldots, d_n$, let $DE(d_1, \ldots, d_n)$ and $DO(d_1, \ldots, d_n)$ denote, respectively, the sets of all even and odd orientations of G in which the out-degree of the vertex v_i is d_i, for $1 \leqslant i \leqslant n$. In this notation, one can check that

$$f_G(x_1, \ldots, x_n) = \sum_{d_1, \ldots, d_n \geqslant 0} \left(\left| DE(d_1, \ldots, d_n) \right| - \left| DO(d_1, \ldots, d_n) \right| \right) \Pi_{i=1}^n x_i^{d_i}.$$

Consider, now, the given orientation D which lies in $DE(d_1, \ldots, d_n) \cup DO(d_1, \ldots, d_n)$. For any orientation $D_2 \in DE(d_1, \ldots, d_n) \cup DO(d_1, \ldots, d_n)$, let $D \oplus D_2$ denote the set of all oriented edges of D whose orientation in D_2 is in the opposite direction. Since the out-degree of every vertex in D is equal to its out-degree in D_2, it follows that $D \oplus D_2$ is an Eulerian subgraph of D. Moreover, $D \oplus D_2$ is even as an Eulerian subgraph if and only if D and D_2 are both even or both odd. The mapping $D_2 \longrightarrow D \oplus D_2$ is clearly a bijection between $DE(d_1, \ldots, d_n) \cup DO(d_1, \ldots, d_n)$ and the set of all Eulerian subgraphs of D. In the case where D is even, it maps even orientations to even (Eulerian) subgraphs, and odd orientations to odd subgraphs. Otherwise, it maps even orientations to odd subgraphs, and odd orientations to even subgraphs. In any case,

$$\left| |DE(d_1, \ldots, d_n)| - |DO(d_1, \ldots, d_n)| \right| = |EE(D) - EO(D)|.$$

Therefore, the absolute value of the coefficient of the monomial $\Pi_{i=1}^{n} x_i^{d_i}$ in the standard representation of $f_G = f_G(x_1, \ldots, x_n)$ as a linear combination of monomials, is $|EE(D) - EO(D)|$. In particular, if $EE(D) \neq EO(D)$, then this coefficient is not zero and the desired result follows from Theorem 1.2. $\qquad \square$

An interesting application of Theorem 7.2 has been obtained by Fleischner and Stiebitz in [29], solving a problem raised by Du, Hsu and Hwang in [23], as well as a strengthening of it suggested by Erdős.

Theorem 7.3 ([29]). *Let G be a graph on $3n$ vertices, whose set of edges is the disjoint union of a Hamilton cycle and n pairwise vertex-disjoint triangles. Then the choice number and the chromatic number of G are both 3.*

The proof is based on a subtle parity argument that shows that, if D is the digraph obtained from G by directing the Hamilton cycle as well as each of the triangles cyclically, then $EE(D) - EO(D) \equiv 2 \pmod 4$. The result thus follows from Theorem 7.2.

Another application of Theorem 7.2 together with some additional combinatorial arguments is the following result, which solves an open problem from [28].

Theorem 7.4 ([13]). *The choice number of every planar bipartite graph is at most 3.*

This is tight, since $\mathrm{ch}(K_{2,4}) = 3$.

Recall that the List Colouring Conjecture (Conjecture 7.1) asserts that $\mathrm{ch}'(G) = \chi'(G)$ for every graph G. In order to try to apply Theorem 7.2 for tackling this problem, it is useful to find a more convenient expression for the difference $EE(D) - EO(D)$, where D is the appropriate orientation of a given line graph. Such an expression is described in [2] for line graphs of d-regular graphs of chromatic index d. This expression is the sum, over all proper d-edge colourings of the graph, of an appropriately defined *sign* of the colouring. See [2] for more details, and [36] for a related discussion. Combining this with a known result of [57] (which asserts that for planar cubic graphs of chromatic index 3 all proper 3-edge colourings have the same sign), and with the Four Colour Theorem, the following result, observed by F. Jaeger and M. Tarsi, follows immediately.

Corollary 7.5. *For every 2-connected cubic planar graph G, $\mathrm{ch}'(G) = 3$.*

Note that the above result is a strengthening of the Four Colour Theorem, which is well known to be equivalent to the fact that the chromatic index of any such graph is 3.

As shown in [25], it is possible to extend this proof to any d-regular planar multigraph with chromatic index d.

Another interesting application of the algebraic method described above appears in [34], where the authors apply it to show that the List Colouring Conjecture holds for complete graphs with an odd number of vertices, and to improve the error term in the asymptotic estimate of Kahn for the maximum possible list chromatic index of a simple graph with maximum degree d. Finally we mention that Galvin [31] recently proved that the List Colouring Conjecture holds for any bipartite multigraph, by an elementary, non-algebraic method.

8. The Permanent Lemma

The following lemma is a slight extension of a lemma proved in [12]. As shown below, it is an immediate corollary of Theorem 1.2 and has several interesting applications.

Lemma 8.1 (The Permanent Lemma).   *Let $A = (a_{ij})$ be an $n \times n$ matrix over a field F, and suppose its permanent $\mathrm{Per}(A)$ is nonzero (over F). Then, for any vector $b = (b_1, b_2, \ldots, b_n) \in F^n$ and for any family of sets $S_1, S_2, \ldots, S_n$ of F, each of cardinality 2, there is a vector $x \in S_1 \times S_2 \times \cdots \times S_n$ such that, for every i, the ith coordinate of Ax differs from b_i.*

Proof.   The polynomial

$$P(x_1, x_2, \ldots, x_n) = \prod_{i=1}^{n} \left[\sum_{j=1}^{n} a_{ij} x_j - b_j \right]$$

is of degree n and the coefficient of $\prod_{i=1}^{n} x_i$ in it is $\mathrm{Per}(A) \neq 0$. The result thus follows from Theorem 1.2.   □

Note that, in the special case $S_i = \{0, 1\}$ for every i, the above lemma asserts that, if the permanent of A is nonzero, then for any vector b there is a subset of the column-vectors of A whose sum differs from b in all coordinates.

A conjecture of Jaeger asserts that, for any field with more than 3 elements and for any nonsingular $n \times n$ matrix A over the field, there is a vector x so that both x and Ax have nonzero coordinates. Note that for the special case of fields of characteristic 2 this follows immediately from the Permanent Lemma. Simply take b to be the zero vector, let each S_i be an arbitrary subset of size 2 of the field that does not contain zero, and observe that in characteristic 2 the permanent and the determinant coincide, implying that $\mathrm{Per}(A) \neq 0$. With slightly more work relying on some simple properties of the permanent function, the conjecture is proved in [12] for every non-prime field. It is still open for prime fields and, in particular, for $p = 5$.

Let $f(n, d)$ denote the minimum possible number f so that every set of f lattice points in the d-dimensional Euclidean space contains a subset of cardinality n whose centroid is also a lattice point. The problem of determining or estimating $f(n, d)$ was suggested by Harborth [35], and studied by various authors.

It is convenient to reformulate the definition of $f(n, d)$ in terms of sequences of elements of the abelian group Z_n^d. In these terms, $f(n, d)$ is the minimum possible f so that every sequence of f members of Z_n^d contains a subsequence of size n, the sum of whose elements (in the group) is 0.

By an old result of Erdős, Ginzburg and Ziv [27], $f(n, 1) = 2n - 1$ for all n. The main part of the proof of this statement is its proof for prime values of $n = p$, as the general case can then be easily proved by induction.

Proposition 8.2 ([27]). *For any prime p, any sequence of $2p - 1$ members of Z_p contains a subsequence of cardinality p, the sum of whose members is 0 (in Z_p).*

Proof. There are many proofs of this result. Here is one using the Permanent Lemma. Given $2p-1$ members of Z_p, renumber them $a_1, a_2, \ldots, a_{2p-1}$ such that $0 \leqslant a_1 \leqslant \cdots \leqslant a_{2p-1}$. If there is an $i \leqslant p - 1$ such that $a_i = a_{i+p-1}$, then $a_i + a_{i+1} + \cdots + a_{i+p-1} = 0$, as needed. Otherwise, let A denote the $(p - 1) \times (p - 1)$ all one matrix, and define $S_i = \{a_i, a_{i+p-1}\}$ for all $1 \leqslant i \leqslant p - 1$. Let $b_1, \ldots, b_{p-1}$ be the set of all elements of Z_p besides $-a_{2p-1}$. Since $\text{Per}(A) = (p - 1)! \neq 0$, by Lemma 8.1, there are $s_i \in S_i$ such that the sum $\sum_{j=1}^{p-1} s_i$ differs from each b_j and is thus equal to $-a_{2p-1}$. Hence, in Z_p,

$$a_{2p-1} + \sum_{i=1}^{p-1} s_i = 0,$$

completing the proof.   □

Kemnitz [40] conjectured that $f(n, 2) = 4n - 3$, observed that $f(n, 2) \geqslant 4n - 3$ for all n and proved his conjecture for $n = 2, 3, 5$ and 7. As in the one-dimensional case, it suffices to prove this conjecture for prime values p. In [5] it is shown that $f(p, 2) \leqslant 6p - 5$ for every prime p. The details are somewhat complicated, but the main tool is again the Permanent Lemma mentioned above.

An *additive basis* in a vector space Z_p^n is a collection C of (not necessarily distinct) vectors, so that for every vector u in Z_p^n there is a subset of C, the sum of whose elements is u. Motivated by the study of universal flows in graphs, Jaeger, Linial, Payan and Tarsi [37] conjectured that for every prime p there exists a constant $c(p)$, such that any union of $c(p)$ *linear* bases of Z_p^n contains an additive basis. This conjecture is still open, but in [9] it is shown that any union of $\lceil (p - 1) \log_e n \rceil + p - 2$ linear bases of Z_p^n contains such an additive basis. Here, too, the Permanent Lemma plays a crucial role in the proof. The main idea is to observe how it can be applied to give equalities rather than inequalities (extending the very simple application described in the proof of Proposition 8.2 above). Here is the basic approach. For a vector v of length n over Z_p, let v^* denote the tensor product of v with the all one vector of length $p - 1$. Thus v^* is a vector of length $(p - 1)n$ obtained by concatenating $(p-1)$ copies of v. In this notation, the following result follows from the Permanent Lemma.

Lemma 8.3. *Let $S = (v_1, v_2, \ldots, v_{(p-1)n})$ be a sequence of $(p-1)n$ vectors of length n over Z_p, and let A be the $(p-1)n \times (p-1)n$ matrix whose columns are the vectors $v_1^*, v_2^*, \ldots, v_{(p-1)n}^*$. If $\mathrm{Per}(A) \neq 0$ (over Z_p), then the sequence S is an additive basis of Z_p^n.*

Proof. For any vector $b = (b_1, b_2, \ldots, b_n)$, let u_b be the concatenation of the $(p-1)$ vectors $b + j, b + 2j, \ldots, b + (p-1)j$, where j is the all one vector of length n. By the Permanent Lemma with all sets $S_i = \{0, 1\}$, there is a subset $I \subset \{1, 2, \ldots, (p-1)n\}$ such that the sum $\sum_{i \in I} v_i^*$ differs from u_b in all coordinates. This supplies $(p-1)$ forbidden values for every coordinate of the sum $\sum_{i \in I} v_i$, and hence implies that $\sum_{i \in I} v_i = b$. Since b was arbitrary, this completes the proof. $\qquad \square$

In [9] it is shown that from any set consisting of all elements in the union of an appropriate number of linear bases of Z_p^n it is possible to choose $(p-1)n$ vectors satisfying the assumptions of the lemma. This is done by applying some properties of the permanent function. The details can be found in [9]. The following conjecture seems plausible, and would imply, if true, that the union of any set of p bases of Z_p^n is an additive basis.

Conjecture 8.4. *For any p nonsingular $n \times n$ matrices $A_1, A_2, \ldots, A_p$ over Z_p, there is an $n \times pn$ matrix C such that the $pn \times pn$ matrix*

$$
M' = \begin{bmatrix}
A_1 & A_2 & \cdots & A_{p-1} & A_p \\
A_1 & A_2 & \cdots & A_{p-1} & A_p \\
\cdot & \cdot & \cdot & \cdot & \cdot \\
\cdot & \cdot & \cdot & \cdot & \cdot \\
A_1 & A_2 & \cdots & A_{p-1} & A_p \\
& & C & &
\end{bmatrix}
$$

has a nonzero permanent over Z_p.

We close this section with a simple result about directed graphs. A *one-regular* subgraph of a digraph is a subgraph of it in which all out-degrees and all in-degrees are precisely 1 (that is, a spanning subgraph which is a union of directed cycles).

Proposition 8.5. *Let $D = (V, E)$ be a digraph containing a one-regular subgraph. Then, for any assignment of a set S_v of two reals for each vertex v of V, there is a choice $c(v) \in S_v$ for every v, so that for every vertex u the sum $\sum_{v:(u,v) \in E} c(v) \neq 0$.*

Proof. Let $A = (a_{u,v})$ be the adjacency matrix of D defined by $a_{u,v} = 1$ if and only if $(u, v) \in E$ and $a_{u,v} = 0$ otherwise. By the assumption, the permanent of A over the reals is strictly positive. The result thus follows from the Permanent Lemma. $\qquad \square$

9. Ideals of polynomials and combinatorial properties

There are several known results that assert that a combinatorial structure satisfies a certain combinatorial property if and only if an appropriate polynomial associated with it lies in a properly defined ideal. Here are three known results of this type, all applying the graph polynomial defined in Section 7.

Theorem 9.1 (Li and Li [41]). *A graph G does not contain an independent set of $k+1$ vertices if and only if the graph polynomial f_G lies in the ideal generated by all graph polynomials of unions of k pairwise vertex disjoint complete graphs that span its set of vertices.*

Theorem 9.2 (Kleitman and Lovász [42, 43]). *A graph G is not k-colourable if and only if the graph polynomial f_G lies in the ideal generated by all graph polynomials of complete graphs on $k+1$ vertices.*

Theorem 9.3 (Alon and Tarsi [13]). *A graph G on the n vertices $\{1, 2, \ldots, n\}$ is not k-colourable if and only if the graph polynomial f_G lies in the ideal generated by the polynomials $x_i^k - 1$, $(1 \leqslant i \leqslant n)$.*

Here is a quick proof of the last theorem, using Theorem 1.1.

Proof of Theorem 9.3. If f_G lies in the ideal generated by the polynomials $x_i^k - 1$ then it vanishes whenever each x_i attains a value which is a kth root of unity. This means that in any colouring of the vertices of G by the kth roots of unity, there is a pair of adjacent vertices that get the same colour, implying that G is not k-colourable.

Conversely, suppose G is not k-colourable. Then f_G vanishes whenever each of the polynomials $g_i(x_i) = x_i^k - 1$ vanishes, and thus, by Theorem 1.1, f_G lies in the ideal generated by these polynomials. $\qquad\square$

In [14] we show that a certain weighted sum over the proper k-colourings of G can be computed in a simple manner from its graph polynomial f_G. In the same note we also claim to provide a short proof of Theorem 9.2, based on the Hajós–Ore Theorem, but as pointed out by Bjarne Toft [55] this proof contains a subtle error. As described in Section 7, there are several interesting combinatorial consequences that can be derived from (some versions of) Theorem 9.3, but even without any consequences, such theorems are interesting in their own. One reason for this is that these theorems characterize *coNP*-complete properties, which, according to the common belief that the complexity classes *NP* and *coNP* differ, cannot be checked by a polynomial time algorithm.

Using Theorem 1.1 it is not difficult to generate results of this type. We illustrate this with two examples, described below. Many other results can be formulated and proved in a similar manner. It would be nice to deduce any interesting combinatorial consequences of these results or their relatives.

The *bandwidth* of a graph $G = (V, E)$ on n vertices is the minimum integer k such that there is a bijection $f : V \mapsto \{1, 2, \ldots, n\}$ satisfying $|f(u) - f(v)| \leqslant k$ for every edge $uv \in E$. This invariant has been studied extensively by various researchers. See, for instance, [20] for a survey.

Proposition 9.4. *The bandwidth of a graph $G = (V, E)$ on a set $V = \{1, 2, \ldots, n\}$ of n vertices is at least $k + 1$ if and only if the polynomial*

$$Q_{G,k}(x_1, \ldots, x_n) = \prod_{1 \leqslant i < j \leqslant n} (x_i - x_j) \prod_{ij \in E, i < j} \prod_{k < |l| < n} (x_i - x_j - l)$$

lies in the ideal generated by the polynomials

$$\left\{ g_i(x_i) = \prod_{j=1}^{n} (x_i - j), \quad 1 \leqslant i \leqslant n \right\}.$$

Proof. If $Q_{G,k}$ lies in the above-mentioned ideal, then it vanishes whenever we substitute a value in $\{1, 2, \ldots, n\}$ for each x_i. In particular, it vanishes when we substitute distinct values for these variables, implying that there is some edge $ij \in E$ for which $|x_i - x_j| > k$, and hence the bandwidth of G exceeds k.

Conversely, assume the bandwidth of G exceeds k. We claim that $Q_{G,k}(x_1, \ldots, x_n)$ vanishes whenever each x_i attains a value in $\{1, 2, \ldots, n\}$. Indeed, if two of the variables attain the same value, the first product $(\prod_{1 \leqslant i < j \leqslant n} (x_i - x_j))$ in the definition of $Q_{G,k}$ vanishes. Otherwise, the numbers x_i form a permutation of the members of $\{1, 2, \ldots, n\}$ and thus, by the assumption on the bandwidth, there is some edge $ij \in E$ for which $|x_i - x_j| > k$, implying that the polynomial vanishes in this case as well. Therefore, $Q_{G,k}$ vanishes whenever each x_i lies in $\{1, 2, \ldots, n\}$ and thus, by Theorem 1.1, it lies in the ideal generated by the polynomials $g_i(x_i)$, completing the proof. $\square$

A *hypergraph* H is a pair (V, E), where V is a finite set, whose elements are called *vertices*, and E is a collection of subsets of V, called *edges*. It is *k-uniform* if each edge contains precisely k vertices. Thus, a 2-uniform hypergraph is simply a graph. H is *2-colourable* if there is a vertex colouring of H with two colours so that no edge is monochromatic.

Proposition 9.5. *The 3-uniform hypergraph $H = (V, E)$ is not 2-colourable if and only if the polynomial*

$$\prod_{e \in E} \left[\left(\sum_{v \in e} x_v \right)^2 - 9 \right]$$

lies in the ideal generated by the polynomials $\{x_v^2 - 1 : v \in V\}$.

Proof. The proof is similar to the previous one. If the polynomial lies in that ideal, then it vanishes whenever each x_v attains a value in $\{-1, 1\}$, implying that some edge is monochromatic in each vertex colouring by $\{-1, 1\}$, and hence implying that H is not 2-colourable. Conversely, if H is not 2-colourable, then in every vertex colouring by the numbers -1 and $+1$ some edge is monochromatic, implying that the polynomial vanishes in each such point, and thus showing, by Theorem 1.1, that it lies in the above ideal. $\square$

Note that, since the properties characterized in any of the theorems in this section are *coNP*-complete, it is possible to use the usual reductions and, for each *coNP*-complete problem, obtain a characterization in terms of some ideals of polynomials. In most cases, however, the known reductions are somewhat complicated, and would thus lead to cumbersome polynomials which are not likely to imply any interesting consequences. The results mentioned here are in terms of relatively simple polynomials, and are therefore more likely to be useful.

10. Concluding remarks

The discussion in Section 7, as well as that in Section 9, raises the hope that the polynomial approach might be helpful in the study of the Four Colour Theorem. This certainly deserves more attention. Further results in the study of the List Colouring Conjecture (Conjecture 7.1) using the algebraic technique are also desirable.

Most proofs presented in this paper are based on the two basic theorems, proved in Section 2, whose proofs are algebraic, and hence non-constructive in the sense that they supply no efficient algorithm for solving the corresponding algorithmic problems.

In the classification of algorithmic problems according to their complexity, it is customary to try and identify the problems that can be solved efficiently, and those that *probably* cannot be solved efficiently. A class of problems that can be solved efficiently is the class P of all problems for which there are deterministic algorithms whose running time is polynomial in the length of the input. A class of problems that probably cannot be solved efficiently are all the NP-complete problems. An extensive list of such problems appears in [32]. It is well known that if any of them can be solved efficiently, then so can all of them, since this would imply that the two complexity classes P and NP are equal.

Is it possible to modify the algebraic proofs given here so that they yield efficient ways of solving the corresponding algorithmic problems? It seems likely that such algorithms do exist. This is related to questions regarding the complexity of search problems that have been studied by several researchers. See, for instance, [38].

In the study of complexity classes like P and NP, one usually considers only decision problems, that is, problems for which the only two possible answers are 'yes' or 'no'. However, the definitions extend easily to the so-called 'search' problems, which are problems where a more elaborate output is sought. The search problems corresponding to the complexity classes P and NP are sometimes denoted by FP and FNP.

Consider, for example, the obvious algorithmic problem suggested by Theorem 6.1 (for $p = 3$, say). Given a simple graph with average degree that exceeds 4 and maximum degree 5, it contains, by this theorem, a 3-regular subgraph. Can we find such a subgraph in polynomial time?

It seems plausible that finding such a subgraph should not be a very difficult task. However, our proof provides no efficient algorithm for accomplishing this task. The situation is similar with many other algorithmic problems corresponding to the various results presented here. Can we, given an input graph satisfying the assumptions of Theorem 7.3 and given a list of three colours for each of its vertices, find, in polynomial time, a proper vertex colouring assigning each vertex a colour from its class? Similarly, can we colour properly the edges of any given planar cubic 2-connected graph using given lists of three colours per edge, in polynomial time?

These problems remain open. Note, however, that any efficient procedure that finds, for a given input polynomial that satisfies the assumptions of Theorem 1.2, a point $(s_1, s_2, \ldots, s_n)$ satisfying its conclusion, would provide efficient algorithms for most of these algorithmic problems. It would thus be interesting to find such an efficient procedure. See also [1] for a related discussion of other algorithmic problems.

Another computational aspect suggested by the results in Section 9 is the complexity of the representation of polynomials in the form that shows they lie in certain ideals. Thus, for example, by Proposition 9.5, a 3-uniform hypergraph is not 2-colourable if and only if the polynomial associated with it in that proposition is a linear combination with polynomial coefficients of the polynomials $x_v^2 - 1$. Since the problem of deciding whether such a given input hypergraph is not 2-colourable is *coNP*-complete, the existence of a representation like this that can be checked in polynomial time would imply that the complexity classes *NP* and *coNP* coincide, and this is believed by most researchers not to be the case.

In this paper we have developed and discussed a technique in which polynomials are applied for deriving combinatorial consequences. There are several other known proof techniques in combinatorics that are based on properties of polynomials. The most common and successful one is based on a dimension argument. This is the method of proving an upper bound for the size of a collection of combinatorial structures satisfying certain prescribed properties by associating each structure with a polynomial in some space of polynomials, showing that these polynomials are linearly independent, and then deducing the required bound from the dimension of the corresponding space. There are many interesting results proved in this manner: see, for instance, [33], [15], [16] and [3] for surveys of results of this type.

References

[1] Alon, N. (1991) Non-constructive proofs in combinatorics. In *Proc. International Congress of Mathematicians, Kyoto 1990, Japan*, Springer, Tokyo, pp. 1421–1429.

[2] Alon, N. (1993) Restricted colorings of graphs. In *Surveys in Combinatorics, Proc. 14th British Combinatorial Conference*, Vol. 187 of *London Math. Soc. Lecture Notes* (K. Walker, ed.), Cambridge University Press, pp. 1–33.

[3] Alon, N. (1995) Tools from higher algebra. In *Handbook of Combinatorics* (R. Graham, M. Grötschel and L. Lovász, eds), Elsevier, pp. 1749–1783.

[4] Alon, N. and Caro, Y. (1993) On three zero-sum Ramsey-type problems. *J. Graph Theory* **17** 177–192.

[5] Alon, N. and Dubiner, M. (1993) Zero-sum sets of prescribed size. In *Combinatorics: Paul Erdős is Eighty*, Vol. 1, *Bolyai Society Math. Studies*, Keszthely, Hungary, pp. 33–50.

[6] Alon, N., Friedland, S. and Kalai, G. (1984) Regular subgraphs of almost regular graphs. *J. Combinatorial Theory Ser. B* **37** 79–91. Also: Alon, N., Friedland, S. and Kalai, G. (1984) Every 4-regular graph plus an edge contains a 3-regular subgraph. *J. Combinatorial Theory Ser. B* **37** 92–93.

[7] Alon, N. and Füredi, Z. (1993) Covering the cube by affine hyperplanes. *European J. Combinatorics* **14** 79–83.

[8] Alon, N., Kleitman, D., Lipton, R., Meshulam, R., Rabin, M. and Spencer, J. (1991), Set systems with no union of cardinality 0 modulo *m*. *Graphs and Combinatorics* **7** 97–99.

[9] Alon, N., Linial, N. and Meshulam, R. (1991) Additive bases of vector spaces over prime fields. *J. Combin. Theory Ser. A* **57** 203–210.

[10] Alon, N., Nathanson, M. B. and Ruzsa, I. Z. (1995) Adding distinct congruence classes modulo a prime. *Amer. Math. Monthly* **102** 250–255.

[11] Alon, N., Nathanson, M. B. and Ruzsa, I. Z. (1996) The polynomial method and restricted sums of congruence classes. *J. Number Theory* **56** 404–417.

[12] Alon, N. and Tarsi, M. (1989) A nowhere-zero point in linear mappings. *Combinatorica* **9** 393–395.

[13] Alon, N. and Tarsi, M. (1992) Colorings and orientations of graphs. *Combinatorica* **12** 125–134.

[14] Alon, N. and Tarsi, M. (1997) A note on graph colorings and graph polynomials. *J. Combin. Theory Ser. B* **70** 197–201.

[15] Babai, L. and Frankl, P. *Linear Algebra Methods in Combinatorics*. To appear.

[16] Blokhuis, A. (1993) Polynomials in finite geometries and combinatorics. In *Surveys in Combinatorics, Proc. 14th British Combinatorial Conference*, Vol. 187 of *London Math. Soc. Lecture Notes* (K. Walker, ed.), Cambridge University Press, pp. 35–52.

[17] Bollobás, B. (1978) *Extremal Graph Theory*, Academic Press.

[18] Bollobás, B. and Harris, A. J. (1985) List colorings of graphs. *Graphs and Combinatorics* **1** 115–127.

[19] Bollobás, B. and Leader, I. (1996) Sums in the grid. *Discrete Math.* **162** 31–48.

[20] Chung, F. R. K. (1988) Labelings of graphs. In *Selected Topics in Graph Theory*, Vol. 3, Academic Press, pp. 151–168.

[21] Davenport, H. (1935) On the addition of residue classes. *J. London Math. Soc.* **10** 30–32.

[22] Dias da Silva, J. A. and Hamidoune, Y. O. (1994) Cyclic spaces for Grassmann derivatives and additive theory. *Bull. London Math. Soc.* **26** 140–146.

[23] Du, D. Z., Hsu, D. F. and Hwang, F. K. (1993) The Hamiltonian property of consecutive-*d* digraphs. *Mathematical and Computer Modelling* **7** 61–63.

[24] Eliahou, S. and Kervaire, M. (1998) Sumsets in vector spaces over finite fields. *J. Number Theory* **71** 12–39.

[25] Ellingham, M. N. and Goddyn, L. (1996) List edge colorings of some 1-factorable multigraphs. *Combinatorica* **16** 343–352.

[26] Erdős, P. and Graham, R. L. (1980) *Old and New Problems and Results in Combinatorial Number Theory*, L'Enseignement Mathématique, Geneva.

[27] Erdős, P., Ginzburg, A. and Ziv, A. (1961) Theorem in the additive number theory. *Bull. Research Council Israel* **10F** 41–43.

[28] Erdős, P., Rubin, A. L. and Taylor, H. (1979) Choosability in graphs. In *Proc. West Coast Conf. on Combinatorics, Graph Theory and Computing. Congressus Numerantium* **XXVI** 125–157.

[29] Fleischner, H. and Stiebitz, M. (1992) A solution to a coloring problem of P. Erdős. *Discrete Math.* **101** 39–48.

[30] Freiman, G. A., Low, L. and Pitman, J. (1993) The proof of Paul Erdős' conjecture of the addition of different residue classes modulo a prime number. In *Structure Theory of Set Addition*, CIRM Marseille, pp. 99–108.

[31] Galvin, F. (1995) The list chromatic index of a bipartite multigraph. *J. Combin. Theory Ser. B* **63** 153–158.

[32] Garey, M. R. and Johnson, D. S. (1979) *Computers and Intractability: A guide to the Theory of NP-Completeness*, W. H. Freeman and Company, New York.

[33] Godsil, C. (1995) Tools from linear algebra. In *Handbook of Combinatorics* (R. Graham, M. Grötschel and L. Lovász, eds), Elsevier, pp. 1705–1748.

[34] Häggkvist, R. and Janssen, J. (1997) New bounds on the list chromatic index of the complete graph and other simple graphs. *Combin. Probab. Comput.* **6** 295–313.

[35] Harborth, H. (1973) Ein Extremalproblem für Gitterpunkte. *J. Reine Angew. Math.* **262/263** 356–360.

[36] Jaeger, F. (1989) On the Penrose number of cubic diagrams. *Discrete Math.* **74** 85–97.

[37] Jaeger, F., Linial, N., Payan, C. and Tarsi, M. (1992) Group connectivity of graphs- a nonhomogeneous analogue of nowhere-zero flow. *J. Combin. Theory Ser. B* **56** 165–182.

[38] Johnson, D. S., Papadimitriou, C. H. and Yannakakis, M. (1988) How easy is local search? *JCSS* **37** 79–100.

[39] Kahn, J. (1996) Asymptotically good list colorings. *J. Combin. Theory Ser. A* **73** 1–59.

[40] Kemnitz, A. (1983) On a lattice point problem. *Ars Combinatoria* **16b** 151–160.

[41] Li, S. Y. R. and Li, W. C. W. (1981) Independence numbers of graphs and generators of ideals. *Combinatorica* **1** 55–61.

[42] Lovász, L. (1982) Bounding the independence number of a graph. In *Bonn Workshop on Combinatorial Optimization* (A. Bachem, M. Grötschel and B. Korte, eds), *Ann. Discrete Math.* **16**, North Holland, Amsterdam, pp. 213–223.

[43] Lovász, L. (1994) Stable sets and polynomials. *Discrete Math.* **124** 137–153.

[44] Mansfield, R. (1981) How many slopes in a polygon? *Israel J. Math.* **39** 265–272.

[45] Macmahon, M. P. A. (1915) *Combinatory Analysis*, Chelsea Publishing Company.

[46] Nathanson, M. B. (1996) *Additive Number Theory: Inverse Theorems and the Geometry of Sumsets*, Springer, New York.

[47] Petersen, J. (1891) Die Theorie der regulären Graphen. *Acta Math.* **15** 193–220.

[48] Pyber, L. (1985) Regular subgraphs of dense graphs. *Combinatorica* **5** 347–349.

[49] Pyber, L., Rödl, V. and Szemerédi, E. (1995) Dense graphs without 3-regular subgraphs. *J. Combin. Theory Ser. B* **63** 41–54.

[50] Rickert, U.-W. (1976) Über eine Vermutung in der additiven Zahlentheorie, PhD thesis, Tech. Univ. Braunschweig.

[51] Rödseth, Ö. J. (1994) Sums of distinct residues mod p. *Acta Arith.* **65** 181–184.

[52] Scheim, D. E. (1974) The number of edge 3-colorings of a planar cubic graph as a permanent. *Discrete Math.* **8** 377–382.

[53] Schmidt, W. (1976) *Equations over Finite Fields: An Elementary Approach*, Vol. 536 of *Lecture Notes in Mathematics*, Springer, Berlin.

[54] Taśkinov, V. A. (1982) Regular subgraphs of regular graphs. *Soviet Math. Dokl.* **26** 37–38.

[55] Toft, B. Private communication, July 1998.

[56] Tsai, S. C. (1996) Lower bounds on representing Boolean functions as polynomials in Z_m. *SIAM J. Discrete Math.* **9** 55–62.

[57] Vigneron, L. (1946) Remarques sur les réseaux cubiques de classe 3 associés au probléme des quatre couleurs. *C. R. Acad. Sc. Paris,* **223** 770–772.

[58] Vizing, V. G. (1964) On an estimate on the chromatic class of a p-graph. *Diskret. Analiz.* **3** 25–30. In Russian.

[59] Vizing, V. G. (1976) Coloring the vertices of a graph in prescribed colors (in Russian), *Diskret. Analiz.* No. 29, *Metody Diskret. Anal. v. Teorii Kodov i Shem* **101** 3–10.

[60] van der Waerden, B. L. (1931) *Modern Algebra*, Julius Springer, Berlin.

[61] Yuzvinsky, S. (1981) Orthogonal pairings of Euclidean spaces. *Michigan Math. J.* **28** 109–119.

Combinatorics, Probability and Computing (1999) 8, 31–43.

Connectedness, Classes and Cycle Index

E. A. BENDER[1], P. J. CAMERON[2],

A. M. ODLYZKO[3] and L. B. RICHMOND[4]

[1] Center for Communications Research,
4320 Westerra Court, San Diego, CA 92121, USA
(e-mail: ed@ccrwest.org)

[2] School of Mathematical Sciences, Queen Mary and Westfield College,
Mile End Road, London E1 4NS, England
(e-mail: p.j.cameron@qmw.ac.uk)

[3] AT&T Bell Laboratories, 600 Mountain Avenue,
Murray Hill, NJ 07974–0636, USA
(e-mail: amo@research.att.com)

[4] Department of Combinatorics and Optimization,
Faculty of Mathematics, University of Waterloo,
Waterloo, Ontario N2L 3G1, Canada
(e-mail: lbrichmo@watdragon.uwaterloo.ca)

This paper begins with the observation that half of all graphs containing no induced path of length 3 are disconnected. We generalize this in several directions. First, we give necessary and sufficient conditions (in terms of generating functions) for the probability of connectedness in a suitable class of graphs to tend to a limit strictly between zero and one. Next we give a general framework in which this and related questions can be posed, involving operations on classes of finite structures. Finally, we discuss briefly an algebra associated with such a class of structures, and give a conjecture about its structure.

1. Introduction

The class of graphs containing no induced path of length 3 has many remarkable properties, stemming from the following well-known observation. Recall that an *induced subgraph* of a graph consists of a subset S of the vertex set together with all edges contained in S.

Proposition 1.1. *Let G be a finite graph with more than one vertex, containing no induced path of length 3. Then G is connected if and only if its complement is disconnected.*

Proof.   It is trivial that the complement of a disconnected graph is connected. Moreover, since P_3 is self-complementary, the property of containing no induced P_3 is self-complementary. So let G be a minimal counterexample: thus, G and $\overline{G}$ are connected but, for any vertex v, either $G - v$ or $\overline{G} - v$ is disconnected. Choose a vertex v and assume, without loss, that $G - v$ is disconnected. Then v is joined to a vertex in each component of $G - v$. Since $\overline{G}$ is connected, there is a vertex x' not joined to v (in G). Let w' be a neighbour of v in the component C of $G - v$ containing x'. Then there is a path from w' to x' in C, and hence an edge wx such that $w \sim v$, $x \not\sim v$. If u is a neighbour of v in a different component of $G - v$, then $\{u, v, w, x\}$ induces P_3, contrary to assumption.   □

The particular view of this result we will take here is that a random P_3-free graph on more than one vertex is connected with probability $\frac{1}{2}$. This leads to the following general question. *In which classes of graphs, having good notions of 'connectedness' and 'induced substructure', does it hold that the probability of connectedness of a random n-vertex graph in the class tends to a limit p, with $0 < p < 1$, as $n \to \infty$?* There are two questions here, since we could take either labelled or unlabelled structures.

One example is the class of forests of rooted trees, where the limiting probability of connectedness is $1/e = 0.3679\ldots$ for the labelled structures, and $1/2.997\ldots = 0.3367\ldots$ for the unlabelled structures. (The latter holds because there is a natural bijection between forests of rooted trees on n vertices, and rooted trees on $n + 1$ vertices; and $2.997\ldots$ is the exponential constant in the asymptotic formula for the number of unlabelled trees: see Otter [11]. For the labelled case, see Rényi [15].)

There is no need to restrict ourselves to graphs. The probability of connectedness of a random N-free poset tends to $(\sqrt{5} - 1)/2$, in both the labelled and the unlabelled case (El-Zahar [4]; see also Stanley [18]). (Incidentally, there is no simple explanation for why the golden ratio appears here, nor for why the answer is the same in the labelled and unlabelled case.)

Contrary to what Proposition 1.1 might suggest, this question turns out to have little to do with the detailed structure of the class, but involves the rate of growth of the number of n-element structures. This will be analysed in Section 2, where we show that a sufficient condition can be expressed in terms of convergence and smoothness properties of the generating function.

The relation between connected and arbitrary structures has several analogues, such as that between partial and total structures, or between reduced and arbitrary structures in a class whose members have a natural 'congruence' relation (where a structure is reduced if the congruence is equality). These are best described in terms of two kinds of composition of classes of finite structures, *multiplication* and *substitution*, defined in Section 3. The behaviour of generating functions under these operations is expressed in terms of a cycle index function, described in Section 4. The final section defines a graded algebra based on a class of structures, and gives a conjecture on this algebra and a structure theorem under additional hypotheses.

2. Convergence and smoothness

Let $\mathcal{A}$ be a class of graphs or other structures which has a notion of 'connectedness'; let $\mathcal{C}$ denote the class of connected structures in $\mathcal{A}$. We assume that every member of $\mathcal{A}$ can be expressed uniquely as a disjoint union of members of $\mathcal{C}$, and that any disjoint union of members of $\mathcal{C}$ is in $\mathcal{A}$. Let c_n and C_n be the numbers of unlabelled and labelled structures in $\mathcal{C}$, and a_n and A_n the corresponding numbers for $\mathcal{A}$. (We assume that $c_0 = C_0 = 0$, $a_0 = A_0 = 1$.) As is usual in enumeration theory, we use exponential generating functions $C(z) = \sum_{n=0}^{\infty} C_n z^n / n!$ and $A(z) = \sum_{n=0}^{\infty} A_n z^n / n!$ for labelled structures, and ordinary generating functions $c(z) = \sum_{n=0}^{\infty} c_n z^n$ and $a(z) = \sum_{n=0}^{\infty} a_n z^n$ for unlabelled structures. (This notation will be used throughout this section.) With our assumptions, we have

$$A(z) = \exp(C(z)), \tag{2.1}$$

$$a(z) = \exp\left(\sum_{n=1}^{\infty} \frac{c(z^n)}{n}\right) = \prod_{n=1}^{\infty} (1 - z^n)^{-c_n}. \tag{2.2}$$

(See Wright [19]; we will derive these well-known equations in Section 5.) Now the probability of connectedness of a random n-element structure in $\mathcal{A}$ is c_n/a_n in the unlabelled case, or C_n/A_n in the labelled case. So the general question is as follows. *What conditions on the sequence (C_n) or (c_n) guarantee that C_n/A_n or c_n/a_n tends to a limit strictly between zero and one as $n \to \infty$, where A_n and a_n are defined by the formulae above?*

Similar questions were first considered by Wright [19], who proved the following.

Theorem 2.1. *If $c_n \geq 0$ for all n, then $c_n/a_n \to 1$ if and only if*

(a) *$c(x)$ has radius of convergence 0, and*

(b) *$\sum_{s=1}^{n-1} h_s h_{n-s} = o(h_n)$, where h_n may be either c_n or a_n.*

The same holds for C_n and A_n.

In Cameron [2], it was conjectured that a necessary and sufficient condition for the probability of connectedness to tend to a limit strictly between zero and one is that the appropriate generating function has finite radius of convergence R and converges at $z = R$, and that its coefficients satisfy some 'smoothness' condition. In this section, we prove a result of this form. First, we observe that the convergence condition is necessary.

Theorem 2.2. *Suppose that, with the above notation, $C(z)$ has finite nonzero radius of convergence R, and that $C(z)$ is unbounded on its circle of convergence. Then*

$$\liminf_{n \to \infty} C_n/A_n = 0.$$

The analogous result holds also for $c(z)$ and $a(z)$.

Proof. Consider the labelled case, and suppose that $C_n > \delta A_n$ for all n, where $\delta > 0$. Then $C(z) > \delta(A(z) - 1)$ for $0 \leq z \leq R$, and so $C(z) > \delta \exp(C(z) - 1)$ as $z \to R$. This is clearly impossible if $C(R)$ is divergent. If $C(R)$ is convergent, then $C(z)$ is uniformly convergent for $|z| = R$.

The argument in the unlabelled case is similar. $\square$

Here is an example. Take a finite alphabet Q, with $|Q| = q$, and let $\mathscr{A}$ consist of all finite words in Q. An induced substructure is taken to be a (not necessarily consecutive) subword. (This example can be recast as a relational structure, where a word on n letters is regarded as n-set that is totally ordered and is partitioned into q subsets corresponding to the elements of Q.) A *Lyndon word* is one that is lexicographically smaller than any proper cyclic shift of itself. Now it can be shown that any word can be expressed uniquely as the concatenation, in lexicographically decreasing order, of Lyndon words. Thus, taking 'disjoint union' to mean concatenation in decreasing order, we are in the general situation of this section. In this case, a_n (the number of words of length n) is equal to q^n, while c_n (the number of Lyndon words of length n) is given by the well-known formula

$$c_n = \frac{1}{n} \sum_{d|n} \mu(d) q^{n/d} \sim \frac{q^n}{n}.$$

We see directly that the radii of convergence of $a(z)$ and $c(z)$ are equal to $1/q$, and that both series diverge at $1/q$; also

$$c_n/a_n \sim 1/n \to 0.$$

Furthermore, because words are totally ordered, we have $C_n = n!c_n$ and $A_n = n!a_n$, so also $C_n/A_n \to 0$.

In addition, some smoothness condition on the coefficients is required to ensure that $\liminf(C_n/A_n) = \limsup(C_n/A_n)$ (or the analogous condition for c_n/a_n). An example to show this was given in Cameron [2]. We give two alternative definitions of 'smoothness' that will work. In the analysis that follows, the arguments in the unlabelled and labelled cases are virtually identical. So we speak of the functions $c(z)$ and $a(z)$, but $C(z)$ and $A(z)$ may be substituted.

The first smoothness condition we consider is *Hayman admissibility*, defined in Hayman [8] or Odlyzko [10]. Rather than give the definition here, we quote a theorem of Hayman that frequently allows an easy proof of Hayman admissibility for generating functions in combinatorial situations.

Theorem 2.3 (Hayman).   *Let $f(z)$ and $g(z)$ be Hayman admissible for $|z| < R$, $R \leqslant \infty$. Let $h(z)$ be analytic in $|z| < R$ and real for real z. Let $p(z)$ be a polynomial with real coefficients.*

   (i) *If the coefficients a_n of the Taylor series of $\exp(p(z))$ are positive for all sufficiently large n, then $\exp(p(z))$ is Hayman admissible for all z.*
   (ii) *$\exp(f(z))$ and $f(z)g(z)$ are Hayman admissible in $|z| < R$.*
   (iii) *If, for some $\eta > 0$ and $R_1 < r < R$,*

$$\max_{|z|=r} |h(z)| = O(f(r)^{1-\eta}),$$

   then $f(z) + h(z)$ is Hayman admissible in $|z| < R$. In particular, $f(z) + p(z)$ is Hayman *admissible in $|z| < R$ and, if the leading coefficient of $p(z)$ is positive, then $p(f(z))$ is Hayman admissible in $|z| < R$.*

Hayman [8] proved the following theorem.

Theorem 2.4. *Let $f(z) = \sum f_n z^n$ be Hayman admissible in $|z| < R$. Let*

$$a(r) = \frac{rf'(r)}{f(r)},$$

$$b(r) = ra'(r) = \frac{rf'(r)}{f(r)} + \frac{r^2 f''(r)}{f(r)} - \left(r\frac{f'(r)}{f(r)}\right)^2.$$

Then

$$f_n \sim (2\pi b(r_n))^{-1/2} f(r_n) r_n^{-n} \text{ as } n \to \infty,$$

where r_n is defined uniquely for sufficiently large n by $a(r_n) = n$. Furthermore, for all $\epsilon > 0$, we have $r_n \to R$, $f(r_n) \to \infty$, and $b(r_n) = o(f(r_n)^\epsilon)$ as $n \to \infty$.

The first result of this section is as follows.

Theorem 2.5. *Let $c(x)$ be the generating function for the unlabelled (or labelled) connected structures in a class and let $0 < R \leq \infty$ be the radius of convergence of $c(x)$. If $\exp(c(x))$ is Hayman admissible, then $c(x) \to \infty$ diverges as $x \to R$ and the probability of connectedness of an unlabelled (or labelled) structure goes to 0 as $n \to \infty$.*

Proof. In the unlabelled case, a_n is at least as large as the coefficient of x^n in $\exp(c(x))$. Hence it suffices to work with $f(x) = \exp(c(x))$ in both the labelled and unlabelled cases and show that $c_n/a_n \to 0$.

The divergence of $f(x)$, and hence $c(x)$, at $x = R$ follows from $f(r_n) \to \infty$ in Theorem 2.4. We have

$$
\begin{aligned}
f_n &\sim (2\pi b(r_n))^{-1/2} \exp(c(r_n)) r_n^{-n} \\
&> \exp((1 - \epsilon)c(r_n)) r_n^{-n} \\
&> Mc(r_n)c(r_n) r_n^{-n} \\
&\geq M \log(f(r_n)) c_n,
\end{aligned}
$$

where the last inequality follows from the fact that a sum of nonnegative terms is at least as large as a single term. Since $f(r_n) \to \infty$, the proof is complete. $\square$

The other smoothness condition we impose on $c(x)$ is satisfaction of the Flajolet–Odlyzko singularity analysis [6]. See [10, Section 11], for the definition of a function of slow variation at ∞, and for the definition of the region Δ. Flajolet and Odlyzko proved the following result.

Theorem 2.6. *Suppose that $c(x)$ has a unique singularity at R on its circle of convergence, that the radius of convergence of $h(x)$ exceeds R, that $L(u)$ is a function of slow variation* *at* ∞, *and that*

$$c(x) - h(x) \sim (R - x)^\alpha L(1/(R - x))$$

as $x \to R$ *in* Δ, *where* α *is not a nonnegative integer. Then*

$$c_n \sim \frac{R^{-n} n^{-\alpha-1} L(n)}{\Gamma(\alpha)}$$

as $n \to \infty$.

Remarks.

(1) A similar result is proved in [6] when α is a nonnegative integer: $\Gamma(\alpha)$ must be replaced by a suitable constant.

(2) If $\alpha < 0$ and

$$L(u) = (\log u)^{\beta_1} (\log \log u)^{\beta_2} \dots,$$

then $\exp(c(x))$ is Hayman admissible (Hayman [8]), and we can use Theorem 2.5.

We now come to the other result of this section.

Theorem 2.7. *Suppose that $C(x)$ satisfies the hypotheses of Theorem 2.6. A necessary and sufficient condition for the probability of connectedness of labelled structures in the class to have a limit strictly between 0 and 1 is that $C(x)$ converge at R.*

Suppose that $c(x)$ satisfies the hypotheses of Theorem 2.6 and has radius of convergence $R < 1$. A necessary and sufficient condition for the probability of connectedness of unlabelled structures in the class to have a limit strictly between 0 and 1 is that $C(x)$ converge at R.

Proof. From Theorem 2.1, we cannot have $c_n/a_n \to 1$ if $R > 0$.

Suppose $c(R)$ diverges (that is, if $\alpha \leq 0$). As in the proof of Theorem 2.5, it suffices to consider $a(x) = \exp(c(x))$. We have

$$
\begin{aligned}
a_n &= [x^n] \exp(c(x)) \\
&\geq [x^n] c(x)^2/2 \\
&\sim r^{-n} n^{-2\alpha-1} L(n)^2/\Gamma(2\alpha),
\end{aligned}
$$

and the result follows.

Suppose now that $c(x)$ is convergent at $x = R$. Then, in Δ, we have

$$c(x) - h(x) \sim (R - x)^\alpha L(1/(R - x)),$$

where $\alpha > 0$ and $h(x)$ is analytic in $|x| < R + \delta$ for some $\delta > 0$. Note that $h(R) = c(R)$. In the labelled case, as $x \to R$ in Δ,

$$a(x) - e^{h(x)} \sim e^{h(R)}(R - x)^\alpha L(1/(R - x)),$$

and we conclude that

$$a_n \sim e^{h(R)} c_n,$$

so that $c_n/a_n \to e^{-h(R)} = e^{-c(R)}$ as $n \to \infty$. The unlabelled case is similar provided that $R < 1$. $\square$

For example, Cayley's theorem for the number T_n of labelled trees on n vertices, combined with Stirling's approximation, shows that

$$T_n/n! \sim Cn^{-5/2}e^n.$$

It is known that $T(z) = \sum T_n z^n/n!$ satisfies the hypotheses of Theorem 2.7, and we conclude that the probability of connectedness of a random forest tends to a limit strictly between zero and one. In fact the limit is $1/\sqrt{e}$ (Rényi [15]).

It remains to prove that various natural classes of graphs satisfy 'smoothness' conditions of the types described above. We pose the following problems. If $\mathscr{C}$ is a class of finite graphs, let $\mathscr{X}(\mathscr{C})$ denote the class of finite graphs containing no induced subgraph isomorphic to a member of $\mathscr{C}$. Note that, if every member of $\mathscr{C}$ is connected, then a graph lies in $\mathscr{X}(\mathscr{C})$ if and only if all its connected components do, so the analysis of this section applies (if the appropriate growth and smoothness conditions can be shown).

(a) Is it true that, if $\mathscr{C}$ is finite, then the probability of connectedness of labelled or unlabelled graphs in $\mathscr{X}(\mathscr{C})$ tends to a limit?

(b) Is it true that P_3 is the only finite connected graph H such that, in $\mathscr{X}(\{H\})$, the limiting probability of connectedness is strictly between 0 and 1?

3. Operations on classes

In this section, we propose a general framework in which a number of questions like the probability of connectedness can be posed and studied. We work in the context of a class $\mathscr{A}$ of finite structures, which is closed under isomorphism and closed under taking induced substructures. Thus $\mathscr{A}$ may be the set of models of a universal theory in a first-order relational language. We allow the language to have infinitely many relation symbols, but require that there are only finitely many n-element structures in $\mathscr{A}$ (up to isomorphism), for each n. As in the preceding section, we let a_n and A_n be the numbers of unlabelled and labelled n-element structures in $\mathscr{A}$, and use the ordinary generating function $\sum_{n=0}^{\infty} a_n z^n$ for the sequence (a_n), and the exponential generating function $\sum_{n=0}^{\infty} A_n z^n/n!$ for (A_n). We assume that $a_0 = A_0 = 1$.

Following Fraïssé [7], the *age* of a countable structure M is the class of all finite structures embeddable in M as induced substructures. Among structures satisfying our assumptions, ages are characterized by the *joint embedding property*: given $A, B \in \mathscr{A}$, there exists $C \in \mathscr{A}$ containing both A and B as induced substructures.

We will frequently make use of the class $\mathscr{S}$ of sets (without any structure). We have $S_n = s_n = 1$ for all n; so $S(z) = \exp(z)$, $s(z) = 1/(1-z)$. Other simple classes are $\mathscr{T}$, the *total orders*, with $t_n = 1$, $T_n = n!$, $t(z) = T(z) = 1/(1-z)$; and the class $\mathscr{P}$ of *permutations*, with $P_n = n!$, $p_n = p(n)$ (the partition function), $P(z) = 1/(1-z)$, $p(z) = \prod_{n=1}^{\infty} 1/(1-z^n)$. (Two permutations are isomorphic if and only if they are conjugate in the symmetric group, so the number of unlabelled permutations of an n-set is the number of partitions of n.)

Now we define two operations on classes of structures as follows. Let $\mathscr{A}, \mathscr{B}$ be classes. Then the operation of *multiplication* produces the class $\mathscr{A} \times \mathscr{B}$, defined as follows: a structure in the class with point set X consists of a partition of X into two parts Y, Z (possibly empty), with an $\mathscr{A}$-structure on Y and a $\mathscr{B}$-structure on Z.

Proposition 3.1. *If $\mathscr{C} = \mathscr{A} \times \mathscr{B}$, then $C(z) = A(z)B(z)$ and $c(z) = a(z)b(z)$.* ☐

For example, if $\mathscr{D}$ is the class of *derangements* (permutations with no fixed points), then $\mathscr{P} = \mathscr{D} \times \mathscr{S}$, from which we obtain the exponential generating function for derangement numbers: $D(z) = 1/((1 - z)\exp(z))$.

The operation of *substitution* of $\mathscr{B}$ into $\mathscr{A}$ produces the class $\mathscr{A}[\mathscr{B} - 1]$, defined as follows: a structure in the class on the point set X consists of a partition of X into an arbitrary number of *non-empty* parts, a $\mathscr{B}$-structure on each part, and an $\mathscr{A}$-structure on the set of parts. The -1 in the notation is intended to suggest that we remove the empty structure from $\mathscr{B}$ before performing the substitution.

Proposition 3.2. *If $\mathscr{C} = \mathscr{A}[\mathscr{B} - 1]$, then $C(z) = A(B(z) - 1)$.* ☐

The function $c(z)$ cannot be determined from $a(z)$ and $b(z)$ alone, as we see in the next section.

For example, if $\mathscr{C}$ denotes the class of *cyclic orders*, then the cycle decomposition of a permutation can be expressed as $\mathscr{P} = \mathscr{S}[\mathscr{C} - 1]$, so that $P(z) = S(C(z) - 1)$, in agreement with the values calculated above for $P(z)$ and $S(z)$ and the fact that $C(z) = 1 - \log(1 - z)$ (from $C_n = (n - 1)!$ for $n \geqslant 1$).

Now let $\mathscr{A}$ be a class of graphs closed under disjoint unions, and let $\mathscr{C}$ be the class of connected graphs in $\mathscr{A}$. We have $\mathscr{A} = \mathscr{S}[\mathscr{C} - 1]$. So the problem of the probability of connectedness is an instance of the following more general problem.

Problem. Suppose that $\mathscr{A}$ is obtained from $\mathscr{B}$ and $\mathscr{C}$ by some operation such as multiplication or substitution. Under what conditions does it hold that C_n/A_n or $\math9_{,}/a_n$ tends to a limit strictly between zero and one?

We give two further examples.

Example 1. If $\mathscr{A} = \mathscr{S} \times \mathscr{C}$, then $\mathscr{A}$-structures can be regarded as *partial $\mathscr{C}$-structures*, consisting of a set with a $\mathscr{C}$-structure on a subset. For example, if $\mathscr{A}$ is a class of graphs closed under adding isolated vertices, and $\mathscr{C}$ is the class of members of $\mathscr{A}$ with no isolated vertices, then this relation holds. So our question would be as follows. *When does it occur that the proportion of partial structures that are total tends to a limit between zero and one?* This is discussed in Cameron [2].

Example 2. Suppose that $\mathscr{A} = \mathscr{C}[\mathscr{S} - 1]$. Then an $\mathscr{A}$-structure carries a natural equivalence relation or *congruence* $\equiv$, such that, if a relation holds for an n-tuple of points, then it remains true if some or all of the points are replaced by equivalent ones. With this interpretation, $\mathscr{C}$ is the class of *reduced* structures, those in which the relation $\equiv$ is just equality. For example, in a suitable class $\mathscr{A}$ of graphs, set $v \equiv w$ if v and w have the same neighbour sets; a graph is reduced if different vertices have different neighbour sets. Now our question is as follows. *When does it occur that the proportion of structures that are reduced tends to a limit between zero and one?*

4. Cycle index

It is possible, following Joyal [9], to define a *cycle index* of a class of structures such that the generating functions for labelled and unlabelled structures defined above are specializations of it. Its behaviour under multiplication and substitution can also be described.

Recall that the *cycle index* $z(g)$ of a permutation g on n letters is the monomial in indeterminates $s_1, \ldots, s_n$ given by

$$z(g) = s_1^{c_1(g)} s_2^{c_2(g)} \ldots,$$

where $c_i(g)$ is the number of cycles of length i in the cycle decomposition of g. If G is a permutation group on n letters, the *cycle index* of G is the average of the cycle indices of its elements:

$$Z(G) = \frac{1}{|G|} \sum_{g \in G} z(g).$$

Now if $\mathscr{G}$ is a class of finite permutation groups, containing only finitely many members of degree n for each n, we define the *cycle index* of $\mathscr{G}$ by $\mathscr{Z}(\mathscr{G}) = \sum_{G \in \mathscr{G}} Z(G)$. (This is a formal power series in infinitely many indeterminates, but the assumption guarantees that each monomial occurs only finitely often in the sum.)

This definition can be extended in two ways. First, let $\mathscr{A}$ be a class of structures, as in the preceding section. We define its *cycle index* to be $\mathscr{Z}(\mathscr{A}) = \mathscr{Z}(\{\mathrm{Aut}(A) : A \in \mathscr{A}\})$, where each unlabelled structure in $\mathscr{A}$ is used once in the sum. Now the generating functions for $\mathscr{A}$ are specializations of the cycle index. If Φ is a formal power series, we let $\Phi(s_i \leftarrow t_i)$ be the result of making the substitution t_i for s_i for all i. Some conditions are required in general in order for this to be well defined. For example, if the t_i are formal power series in one indeterminate z, it suffices that t_i has no term of degree less than i.

Proposition 4.1.

(i) $A(z) = \mathscr{Z}(\mathscr{A})(s_i \leftarrow z^i)$.
(ii) $a(z) = \mathscr{Z}(\mathscr{A})(s_1 \leftarrow z, s_i \leftarrow 0 \text{ for } i > 1)$. $\square$

The cycle index behaves as follows under multiplication and substitution.

Proposition 4.2.

(i) $\mathscr{Z}(\mathscr{A} \times \mathscr{B}) = \mathscr{Z}(\mathscr{A})\mathscr{Z}(\mathscr{B})$.
(ii) $\mathscr{Z}(\mathscr{A}[\mathscr{B} - 1]) = \mathscr{Z}(\mathscr{A})(s_i \leftarrow t_i - 1)$, where $t_i = \mathscr{Z}(\mathscr{B})(s_j \leftarrow s_{ij})$. $\square$

From the second part of this proposition, we obtain the missing formula for the generating function for unlabelled structures in $\mathscr{A}[\mathscr{B} - 1]$.

Proposition 4.3. *If* $\mathscr{C} = \mathscr{A}[\mathscr{B} - 1]$, *then* $c(z) = \mathscr{Z}(\mathscr{A})(s_i \leftarrow b(z^i) - 1)$.   □

For the class $\mathscr{S}$ we have

$$\mathscr{Z}(\mathscr{S}) = \exp\left(\sum_{n=1}^{\infty} \frac{s_n}{n}\right).$$

From this, the formula of Section 2 relating connected and arbitrary graphs in a class closed under taking disjoint unions follows.

The other extension is to a class of infinite permutation groups. A permutation group G on a set X is called *oligomorphic* if the number of orbits of G on the set of n-tuples of points of X is finite for all n. See Cameron [1] for an account of these groups.

Clearly, the previous definition of the cycle index of an infinite permutation group makes no sense. However, an oligomorphic permutation group G has a so-called *modified cycle index* $\tilde{Z}(G)$, defined as follows. Choose representatives for the orbits of G on finite subsets of X. (By assumption, there are only finitely many of each size.) For each representative Y, let $G(Y)$ be the permutation group induced on Y by its setwise stabilizer in G. Then $\tilde{Z}(G)$ is defined to be the cycle index of this collection of finite permutation groups.

The relationship with the preceding is as follows. A structure M is said to be *homogeneous* if any isomorphism between finite substructures of M extends to an automorphism of M. Examples of homogeneous structures include the pentagon, the rational numbers $\mathbb{Q}$ (as ordered set), and the random graph or Rado's graph [5, 14]. A theorem of Fraïssé [7] characterizes the ages of countable homogeneous structures. In particular, a class $\mathscr{A}$ of structures satisfying our conditions (that is, closed under isomorphism and under induced substructures and containing only finitely many n-element structures up to isomorphism) is the age of a countable homogeneous structure if and only if it satisfies the *amalgamation property*: if $B, C \in \mathscr{A}$ have isomorphic substructures A, A' respectively, then there is a structure $D \in \mathscr{A}$ in which B and C can be embedded in such a way that the substructures are identified or 'glued together' according to the isomorphism.

Proposition 4.4. *Let M be a countable homogeneous structure with automorphism group G and age $\mathscr{A}$. Then*

(i) *the number of orbits of G on n-element subsets is equal to the number of unlabelled n-element structures in $\mathscr{A}$*

(ii) *the number of orbits of G on n-tuples of distinct elements is equal to the number of labelled n-element structures in $\mathscr{A}$*

(iii) $\tilde{Z}(G) = \mathscr{Z}(\mathscr{A})$.   □

This result gives a convenient translation between ages satisfying the amalgamation property and oligomorphic permutation groups. Under this translation, multiplication and substitution of ages correspond to the direct product (in its intransitive action) and the wreath product (in its imprimitive action) of permutation groups.

5. Algebras

In this section, a graded algebra is associated with a class of finite structures. We make a conjecture about its structure, and give a structure theorem under additional hypotheses. See Cameron [3] for more details of the latter.

Let $\mathscr{A}$ be a class of finite structures satisfying our usual conditions (closed under isomorphism and under induced substructures, and containing only finitely many n-element structures up to isomorphism). For each n, let V_n denote the $\mathbb{Q}$-vector space of all isomorphism-invariant rational functions on the set of n-element structures in $\mathscr{A}$. Thus, $\dim(V_n) = a_n$; a basis for V_n consists of the characteristic functions of the isomorphism classes of n-element structures.

We define

$$\mathrm{Alg}(\mathscr{A}) = \bigoplus_{n=0}^{\infty} V_n,$$

and define a product as follows. Take $f \in V_n$, $g \in V_m$. Then fg is the function in V_{n+m} whose value on the $(n+m)$-element structure $X \in \mathscr{A}$ is given by

$$(fg)(X) = \sum_{\substack{Y \subseteq X \\ |Y|=n}} f(Y)g(X \setminus Y).$$

This multiplication is extended linearly to the whole of $\mathrm{Alg}(\mathscr{A})$. The algebra is easily seen to be commutative and associative. An element of $\mathrm{Alg}(\mathscr{A})$ is said to be *homogeneous of degree* n if it is contained in V_n.

The construction behaves well with respect to multiplication of classes.

Proposition 5.1. $\qquad \mathrm{Alg}(\mathscr{A} \times \mathscr{B}) = \mathrm{Alg}(\mathscr{A}) \otimes_{\mathbb{Q}} \mathrm{Alg}(\mathscr{B}).$ $\qquad\qquad \square$

For $\mathscr{A} = \mathscr{S}$, we have $\dim(V_n) = 1$ for all n, and in fact $\mathrm{Alg}(\mathscr{S})$ is a polynomial algebra in one variable, the generator being the function in V_1 taking the value 1 on all singleton sets.

Conjecture. Suppose that $\mathscr{A}$ has the following property: for any $A, B \in \mathscr{A}$, there exists $C \in \mathscr{A}$ in which A and B can be embedded as disjoint substructures. Then $\mathrm{Alg}(\mathscr{A})$ is an integral domain (*i.e.*, has no divisors of zero).

We make some remarks about this conjecture. First, the condition is clearly necessary. For, if A and B cannot be embedded disjointly in any $\mathscr{A}$-structure, then $f_A f_B = 0$, where f_A is the characteristic function of the isomorphism class of A. Also, the condition is a strengthening of the joint embedding property, so a class $\mathscr{A}$ satisfying it is the age of a countably infinite structure M. In this case, $\mathrm{Alg}(\mathscr{A})$ is a subalgebra of the *reduced incidence algebra* of the poset of finite subsets of M (Rota [17]).

If a graded algebra is a polynomial algebra generated by homogeneous elements, then the relation between the sequence enumerating the polynomial generators by degree and the sequence of dimensions of the homogeneous components is identical to the relation between the sequences enumerating unlabelled connected and arbitrary structures, met with at the start of Section 2. This observation motivates the following result, taken from Cameron [3].

Theorem 5.2. *Let $\mathscr{A}$ be a class of structures. Suppose that $\mathscr{A}$ possesses*

   (i) *a subclass of 'connected' structures*
  (ii) *a partial order $\leqslant$ of 'involvement' on the set of n-element structures for each n*
 (iii) *a commutative and associative 'composition' $\circ$ such that $|A \circ B| = |A| + |B|$.*

Assume that:

   (i) *any structure in $\mathscr{A}$ is uniquely the composition of connected structures*
  (ii) *if $A \in \mathscr{A}$ is partitioned into substructures $A_1, A_2, \ldots$, then $A_1 \circ A_2 \circ \cdots \leqslant A$.*

Then $\mathrm{Alg}(\mathscr{A})$ is a polynomial algebra generated by the characteristic functions of the isomorphism classes of connected structures in $\mathscr{A}$.

The conditions of the theorem are satisfied when $\mathscr{A}$ is the class of all graphs; we take 'connected' to have its usual meaning, 'involvement' to mean 'spanning subgraph', and 'composition' to be 'disjoint union'. A more unusual example involves the words considered in Section 2. As there, a word is 'connected' if it is a Lyndon word (smaller than any proper cyclic shift of itself); 'involvement' is lexicographic order, reversed; and 'composition' is concatenation in lexicographically decreasing order. Now the hypotheses of the theorem are satisfied. The algebra $\mathrm{Alg}(\mathscr{A})$ is the *shuffle algebra*, which occurs in the theory of free Lie algebras (see Reutenauer [16]), and which was shown to be a polynomial algebra by Radford [13].

References

[1]   Cameron, P. J. (1990) *Oligomorphic Permutation Groups*, Vol. 152 of *London Math. Soc. Lecture Notes*, Cambridge University Press, Cambridge.

[2]   Cameron, P. J. (1997) On the probability of connectedness. *Discrete Math.* **167/168** 173–185.

[3]   Cameron, P. J. (1997) The algebra of an age. In *Model Theory of Groups and Automorphism Groups* (D. M. Evans, ed.), Vol. 244 of *London Math. Soc. Lecture Notes*, Cambridge University Press, Cambridge, pp. 126–133.

[4]   El-Zahar, M. H. (1989) Enumeration of ordered sets. In *Algorithms and Order* (I. Rival, ed.), Kluwer, Dordrecht, pp. 327–352.

[5]   Erdős, P. and Rényi, A. (1963) Asymmetric graphs. *Acta Math. Acad. Sci. Hungar.* **14** 295–315.

[6]   Flajolet, P. and Odlyzko, A. M. (1990) Singularity analysis of generating function. *SIAM J. Discrete Math.* **3** 216–240.

[7]   Fraïssé, R. (1953) Sur certains relations qui généralisent l'ordre des nombres rationnels. *C. R. Acad. Sci. Paris* **237** 540–542.

[8]   Hayman, W. K. (1956) A generalisation of Stirling's formula. *J. Reine Angew. Math.* **196** 67–95.

[9]   Joyal, A. (1981) Une théorie combinatoire des séries formelles. *Advances Math.* **42** 1–82.

[10] Odlyzko, A. M. (1995) Asymptotic enumeration methods. In *Handbook of Combinatorics* (R. Graham, M. Grötschel and L. Lovász, eds), Elsevier, Amsterdam, pp. 1063–1229.

[11] Otter, R. (1948) The number of trees. *Ann. Math.* **49** 583–599.

[12] Pólya, G. (1937) Kombinatorische Anzahlbestimmungen für Gruppen, Graphen und chemische Verbindungen. *Acta Math.* **68** 145–254.

[13] Radford, D. E. (1979) A natural ring basis for the shuffle algebra and an application to group schemes. *J. Algebra* **58** 432–454.

[14] Rado, R. (1964) Universal graphs and universal functions. *Acta Arith.* **9** 331–340.

[15] Rényi, A. (1959) Some remarks on the theory of trees. *Publ. Math. Inst. Hungar. Acad. Sci.* **4** 73–85.

[16] Reutenauer, C. (1994) *Free Lie Algebras*, Oxford University Press, Oxford.

[17] Rota, G.-C. (1964) On the foundations of combinatorial theory, I: Theory of Möbius functions. *Z. Wahrscheinlichkeitstheorie* **2** 340–368.

[18] Stanley, R. P. (1974) Enumeration of posets generated by disjoint unions and ordinal sums. *Proc. Amer. Math. Soc.* **45** 295–299.

[19] Wright, E. M. (1967/8) A relationship between two sequences. Part I, *Proc. London Math. Soc.* **17** (1967) 296–304; Part II, *ibid.* **17** (1967) 547–552; Part III, *J. London Math. Soc.* **43** (1968) 720–724.

Combinatorics, Probability and Computing (1999) 8, 45–93.
© 1999 Cambridge University Press

A Tutte Polynomial for Coloured Graphs

BÉLA BOLLOBÁS[1,2] and OLIVER RIORDAN[1]

[1] Department of Mathematical Sciences,
University of Memphis, Memphis TN 38152, USA
(e-mail: bollobas@msci.memphis.edu)

[1] Department of Pure Mathematics and Mathematical Statistics,
University of Cambridge, 16 Mill Lane, Cambridge CB2 1SB, England
(e-mail: B.Bollobas@dpmms.cam.ac.uk)

[2] Institute for Advanced Study,
Olden Lane, Princeton NJ 08540, USA

We define a polynomial W on graphs with colours on the edges, by generalizing the spanning tree expansion of the Tutte polynomial as far as possible: we give necessary and sufficient conditions on the edge weights for this expansion not to depend on the order used. We give a contraction-deletion formula for W analogous to that for the Tutte polynomial, and show that any coloured graph invariant satisfying such a formula can be obtained from W. In particular, we show that generalizations of the Tutte polynomial obtained from its rank generating function formulation, or from a random cluster model, can be obtained from W. Finally, we find the most general conditions under which W gives rise to a link invariant, and give as examples the one-variable Jones polynomial, and an invariant taking values in $\mathbb{Z}/22\mathbb{Z}$.

1. Introduction

1.1. Basic definitions

Throughout this paper, we shall consider the set $\mathscr{G}$ of finite multigraphs, with loops allowed. Usually, we shall call an element of $\mathscr{G}$ a *graph*, but sometimes we shall write *multigraph* for emphasis.

The *Tutte polynomial*, or *dichromate* of [34], is an isomorphism-invariant function $T : \mathscr{G} \longrightarrow \mathbb{Z}[x, y]$, which arises in many different ways. We shall consider four different definitions of the Tutte polynomial.

The first definition, due to Tutte [34], is based on a *spanning tree expansion*. We take an order ϕ on $E(G)$, and, for each spanning tree S of G, use ϕ and S to classify the edges of G into four types. We then assign a weight to each edge, depending on its type, and multiply these weights to find the weight of S. Finally, the Tutte polynomial of G is obtained by summing over all spanning trees S of G. This definition is described precisely in Section 2. Tutte proved the rather surprising result that the polynomial obtained is independent of the order ϕ, and hence defines a graph invariant.

The second definition we shall consider is in terms of contraction-deletion formulae. We shall say that an edge e of G is *ordinary* if e is neither a bridge nor a loop. The Tutte polynomial $T(G; x, y)$ can then be defined by the recurrence relations

$$T(G; x, y) = \begin{cases} xT(G/e; x, y) & \text{if } e \text{ is a bridge,} \\ yT(G - e; x, y) & \text{if } e \text{ is a loop,} \\ T(G - e; x, y) + T(G/e; x, y) & \text{if } e \text{ is ordinary,} \end{cases} \quad (1.1)$$

together with the condition

$$T(E_n) = 1, \quad (1.2)$$

where E_n is the graph with n vertices and no edges. Note that some work is required to show that these conditions do have a solution. One approach is to take an alternative definition of T, for instance via spanning trees, and show that it satisfies (1.1). We give a more direct approach in Section 2.

The third definition we consider is essentially equivalent to the second. We shall call a graph G *basic* if G has no ordinary edges. We can define T by the single relation

$$T(G; x, y) = T(G - e; x, y) + T(G/e; x, y), \quad (1.3)$$

for every ordinary edge e of a graph G, together with the boundary condition

$$T(G; x, y) = x^b y^l, \quad (1.4)$$

for any basic graph G consisting of b bridges and l loops.

The fourth and final definition we shall consider is via the *Whitney–Tutte dichromatic polynomial* [33],

$$Q(G; t, z) = \sum_{H \subseteq G} t^{k(H)} z^{n(H)}, \quad (1.5)$$

where the sum is over all spanning subgraphs H of G, $k(H)$ is the number of components of H, and $n(H)$ is the nullity of H. The Tutte polynomial is then given by

$$T(G; x, y) = (x - 1)^{-k(G)} Q(G; x - 1, y - 1).$$

We shall discuss the relationships between the above definitions in Section 2. In this paper we consider generalizations of the Tutte polynomial to coloured graphs. To be specific, by a *coloured graph* we mean a pair (G, c), where G is a multigraph, and c is a function from $E(G)$ to an arbitrary set Λ of colours.

1.2. Earlier work

Before turning to coloured graphs, it is natural to ask whether the relations (1.1) can be generalized on uncoloured graphs.

In fact, a slight extension of a result of Oxley and Welsh [22] shows that there is a unique map $U : \mathscr{G} \to \mathbb{Z}[x, y, \alpha, \sigma, \tau]$ such that

$$U(E_n) = U(E_n; x, y, \alpha, \sigma, \tau) = \alpha^n$$

for every $n \geqslant 1$, and for every $e \in E(G)$ we have

$$U(G) = \begin{cases} xU(G/e) & \text{if } e \text{ is a bridge,} \\ yU(G-e) & \text{if } e \text{ is a loop,} \\ \sigma U(G-e) + \tau U(G/e) & \text{if } e \text{ is ordinary.} \end{cases} \tag{1.6}$$

Furthermore,

$$U(G) = \alpha^{k(G)} \sigma^{n(G)} \tau^{r(G)} T(G; x/\tau, y/\sigma). \tag{1.7}$$

This answers the question of how far the recurrence relations (1.1) can be extended on uncoloured graphs by introducing new coefficients. This result may be perceived as somewhat negative, in that the function obtained is no more general than the Tutte polynomial. However, the result is still important for two reasons. The first is that it shows that this approach does not give rise to a more general invariant. The second is that it illustrates the generality and importance of the Tutte polynomial, showing that any graph invariant satisfying contraction-deletion relations of this form may be read out of the Tutte polynomial. Particular examples are the chromatic polynomial, the flow polynomial, the number of spanning trees or forests, and the number of connected subgraphs. Many more are given by Welsh [36]. This fact illustrates why the Tutte polynomial is such an important graph invariant.

Returning to coloured graphs, the invariants we are interested in are those which are like the Tutte polynomial in one of two senses. Either they should obey recurrence relations analogous to (1.6), but with each of the variables x, y, σ, τ replaced by one variable for each colour, or they should have a spanning tree expansion analogous to that of the Tutte polynomial. In fact, we shall see in Section 3.1 that these conditions are equivalent.

The first such invariants were defined for *weighted graphs*, where the set Λ of colours is just the set of real numbers. The most direct way to obtain such invariants is to take the rank generating formulation (1.5) of the Tutte polynomial, and add coefficients depending on the weights. For example, for a graph G with weights $w(e) \in \mathbb{R}$ on the edges, Traldi [32] defined the dichromatic polynomial $Q(G; t, z)$ to be

$$Q(G; t, z) = \sum_{H \subseteq G} \left(\prod_{e \in E(G)} w(e) \right) t^{k(H)} z^{n(H)}. \tag{1.8}$$

Traldi showed that this polynomial obeys recurrence relations as above, with the coefficients certain functions of the weights, t and z, and that the polynomial has a spanning tree expansion. In fact, as remarked at the end of [32], this polynomial can be considered as a special case of the random cluster model of Fortuin and Kasteleyn [7], which we describe precisely in Section 4. Again, this comes from an explicit definition, similar to (1.8), which can be thought of in terms of the number of components of a random subgraph of G

The next group of coloured Tutte polynomials were defined for signed graphs, where $\Lambda = \{+, -\}$, motivated by the close connection between signed plane graphs and link diagrams. These started from the very simple description of the Jones polynomial given by Kauffman [13], in terms of state models. This was translated by Thistlethwaite [29] to a single variable signed graph polynomial defined by a spanning tree expansion, and extended to a three variable polynomial by Kauffman [14]. The latter polynomial specializes to the Tutte polynomial on graphs with constant sign. On signed plane graphs, considered as link diagrams, it is exactly the original Kauffman state model, before the conditions guaranteeing Reidemeister invariance are imposed. Kauffman showed that this polynomial satisfies recurrence relations as described above, with the same coefficients as in the spanning tree expansion. Motivated by the connection with link diagrams, the coefficients for contracting a positive edge and deleting a negative edge, for example, were taken to be the same, since these operations on graphs correspond to the same operation on link diagrams.

Also for signed graphs, Murasugi [20] used a rank generating function formulation to define an invariant that is essentially the special case of Traldi's dichromatic polynomial given by taking the weights of positive and negative edges to be x and x^{-1} respectively.

In the papers mentioned above, general polynomials satisfying certain conditions were produced. However, no attempt was made to find the *most* general invariant satisfying these conditions, or to prove that any such invariant must be of a certain form. First Zaslavsky [40] and then Schwärzler and Welsh [26] set themselves exactly such tasks, for very different conditions. Zaslavsky considered invariants satisfying recurrence relations similar to (1.6), with arbitrary coefficients in an arbitrary field K. In fact, the form of the relations he considered is closer to the third definition of the Tutte polynomial, from (1.3) and (1.4), but this makes no difference, as explained in the next section. Zaslavsky also considered matroids rather than graphs, but, as we shall see, this too makes very little difference. He showed that all such invariants are of one of seven types, and described exactly the possible coefficients. He also showed that such invariants have spanning tree expansions.

For signed graphs, Schwärzler and Welsh [26] started from relations with arbitrary coefficients, this time in a ring. However, in their quest for the most general invariant satisfying such relations they make some unstated assumptions. As a result, their description of the most general invariant (Proposition 2.1 in [26]) is incorrect. In particular, it contradicts Theorem 6; for more detail see Remark 5 following this result. Nevertheless, Schwärzler and Welsh obtain an invariant that includes both Kauffman's Tutte polynomial for signed graphs, and Murasugi's polynomial. They also obtain the Jones polynomial by making appropriate substitutions and then checking Reidemeister invariance.

1.3. Summary of results

In this paper we consider two very precise questions, asking for the most general coloured graph invariant satisfying certain conditions. Such questions are important even if the answers turn out to give nothing new, since we then know that a certain approach goes so far, and no further. In fact, we do obtain a new invariant W, which, as an automatic consequence of our approach, includes the previous generalizations as special cases.

The first condition we consider is that of having a spanning tree expansion. We start by taking independent variables for each edge type (defined with respect to an order ϕ and a spanning tree T) and colour, obtaining a function $W(G, c, \phi)$ depending both on the coloured graph (G, c) and on the order ϕ on $E(G)$. We then determine the minimum conditions that must be imposed on these variables to make this invariant independent of the order ϕ. More precisely, we define an order-independent invariant $W(G, c)$ taking values in a quotient ring, such that any other invariant with a spanning tree expansion can be obtained by composing $W(G, c)$ with a suitable ring homomorphism. This invariant retains many properties of the Tutte polynomial. In particular, on graphs whose edges are all the same colour, it essentially coincides with the Tutte polynomial.

The second question we consider is that of satisfying recurrence relations analogous to (1.6), but with different coefficients for edges of different colours. Answering this question will give a coloured graph invariant with an analogous property to that of the function U described above – it will include *any* coloured graph invariant satisfying recurrence relations of this form.

Somewhat surprisingly, it turns out that we already have the invariant we are looking for – it is the same function W as above. Of course, we must do some work to show this. We show on the one hand that W obeys relations of this form, with certain constraints on the variables. On the other hand, we show that these constraints are necessary, in the sense that, unless they are satisfied, there is no invariant satisfying the recurrence relations. This shows that W is the universal invariant we are seeking, and may be used to calculate any coloured graph invariant satisfying recurrence relations of this form, no matter what the coefficients. This includes all the previous generalizations of the Tutte polynomial to coloured, weighted or signed graphs mentioned above.

At this point we note that, like the Tutte polynomial, the invariant W depends only on the cycle matroid of a graph G, and that W can be extended to all matroids. This is because W is determined by the sets of edges that include a cycle, and we only use properties of such sets which hold for general matroids. Having extended W to matroids, we can compare our results with those of Zaslavsky [40]; by not imposing the restriction that the coefficients lie in a field, we obtain one universal invariant, from which the seven classes of invariant described by Zaslavsky can be obtained. This demonstrates the advantages of our unrestricted approach.

As an application, we use W to find the most general link invariant that can be defined via a spanning tree expansion on signed plane graphs. Again, our aim is to make no unnecessary assumptions. Thus, even if we do not obtain a new invariant, we shall have proved that what we do get is the most general link invariant satisfying these conditions. To do this we must first describe exactly the relationship between link diagrams and signed plane graphs, and the exact graph equivalents of Reidemeister moves. This is simple in principle, but there are some complications concerning diagrams with more than one component, which we have not seen described in print. These complications cannot be avoided, as they arise naturally in the expansion considered. Using this correspondence, we obtain an invariant taking values in a certain quotient of $\mathbb{Z}[A, B, d]$, which, since the Jones polynomial has a spanning tree expansion [29], can be used to calculate the Jones polynomial. There seems no reason why this invariant should not distinguish some pairs of links with the same Jones polynomial, but, for the few pairs we have tried so far, it does not.

We remark that all the problems considered here have a property that seems to arise frequently when finding necessary and sufficient conditions for something. This is that, in all cases, the stated conditions are trivially necessary, and what is surprising, and requires some work to prove, is that they are also sufficient. Of course, it is easy to find necessary conditions. The key is to find the *right* conditions, so that they are not too numerous, but they are also sufficient.

The rest of the paper is organized as follows. In Section 2, we give some background concerning the Tutte polynomial, relating the various different definitions considered above. In Section 3, we define a polynomial W on coloured graphs using a spanning tree expansion, keeping the maximum generality allowed by independence of the order used. We then consider recurrence relations similar to (1.6), and show that the most general solution is given by the same function W. In Section 4, we consider the polynomial Z defined by Fortuin and Kasteleyn [7] by considering the number of components of a random subgraph of a graph G. Schwärzler and Welsh [26] showed that their invariant, defined for graphs with only two colours on the edges, can be used to recover a special case of the polynomial Z. Here we show that from W we can recover Z in its full generality. In Section 5, we consider a generalization of the rank generating function, given by Traldi [32], and show that it can also be obtained from W.

In Section 6, we turn to links, establishing the most general conditions under which W is well defined on link diagrams. We give two specific examples of link invariants which can be obtained from W: the Jones polynomial, and a function taking values in $\mathbb{Z}/22\mathbb{Z}$. We also show that the signed graph polynomial defined by Kauffman [14] from his bracket is less general than W restricted to signed graphs.

At this point we shall start again from the beginning, describing the original development of the Tutte polynomial, and the relationships between its various different definitions.

2. The Tutte polynomial

In 1912 Birkhoff [3] proved a determinant formula for the number $p_G(\lambda)$ of (proper) vertex colourings of a graph G with λ distinguishable colours, $1, 2, \ldots, \lambda$, say. He proved also that if G has m_{ij} spanning subgraphs of rank i and nullity j then

$$p_G(\lambda) = \sum_{i,j} (-1)^{i+j} m_{ij} \lambda^{|G|-i}. \tag{2.1}$$

Although our terminology and notation are standard, let us point out that we write $|G|$ for the order of G, $e(G)$ for its size and $k(G)$ for the number of its components; the *rank* of G is $r(G) = |G| - k(G)$, and its *nullity* is $n(G) = e(G) - |G| + k(G) = e(G) - r(G)$. Thus, in particular, $\sum_{i+j=\ell} m_{ij} = \binom{e(G)}{\ell}$.

The investigation of the coefficients m_{ij} was continued by Whitney in his thesis at Harvard, the results of which were published in several papers, including [38] and [39]. (In [38] Whitney rediscovered (2.1) as well.) Let

$$m_i = \sum_j (-1)^j m_{ij},$$

so that

$$p_G(\lambda) = \sum_i (-1)^i m_i \lambda^{|G|-i}.$$

Whitney [39] proved that m_i is the number of i-sets of edges containing no broken cycles. A *broken cycle* is defined in terms of a total (linear) order on $E = E(G)$: it is the edge set of a cycle from which the last edge has been deleted. Note that, *a priori*, m_i depends on which order we choose on E, but the result implies that it is, in fact, independent of the order.

Let $G_1, G_2, \ldots$ be an enumeration of all (isomorphism classes of finite) nonseparable graphs. Another result of Whitney is that $m_{ij} = m_{ij}(G)$ is a polynomial in the numbers $N_1, N_2, \ldots$, where N_h is the number of subgraphs of G isomorphic to G_h.

Also,

$$m_{ij}(G_1 \cup G_2) = \sum_{p,q} m_{pq}(G_1) m_{i-p,j-q}(G_2).$$

Even more importantly, in [39] Whitney remarked that R. M. Foster used the following formula to compute $m_{ij}(G)$: if $e \in E(G)$, then

$$m_{ij}(G) = \begin{cases} m_{ij}(G-e) + m_{i-1,j}(G/e) & \text{if } e \text{ is not a loop,} \\ m_{ij}(G-e) + m_{i,j-1}(G-e) & \text{if } e \text{ is a loop.} \end{cases} \tag{2.2}$$

Here $G - e$ is the graph obtained from G by *deleting* e, and G/e is obtained from G by *contracting* e. Formula (2.2) is, perhaps, the first occurrence of a *contraction-deletion formula*; formulae of this type are of the utmost importance in the study of polynomials related to the Tutte polynomial.

The use of contraction-deletion formulae practically compels us to consider graphs with *loops and multiple edges*, and this is precisely what we shall do. As stated in the Introduction, we shall write $\mathscr{G}$ for the set of (isomorphism classes of finite) graphs with loops and multiple edges. We shall be interested in functions of graphs that are constant on isomorphism classes; that is, we wish to study functions defined on $\mathscr{G}$. Note that for $G \in \mathscr{G}$, $E(G/e)$ is naturally identified with $E(G - e) = E(G) \backslash \{e\}$.

With hindsight it is clear that the two-dimensional array $(m_{ij}(G))_{i,j}$ is best studied by taking the m_{ij} to be the coefficients of a polynomial in two variables, but this important step was taken only several years later, in 1947. Let $q_{k\ell} = q_{k\ell}(G)$ be the number of spanning subgraphs of G with k components and nullity ℓ. Then the *dichromatic polynomial Q* of the graph G can be written as

$$Q(G; t, z) = \sum_{k,\ell} q_{k\ell} t^k z^\ell,$$

so that Q maps $\mathscr{G}$ into $\mathbb{Z}[t, z]$. The polynomial Q is frequently called the *Whitney–Tutte dichromatic polynomial*. It makes little difference whether we use the coefficients $q_{k\ell}$ or m_{ij}, since we have $m_{ij}(G) = q_{|G|-i,j}(G)$. Thus, defining the *rank generating function R* of the graph G as

$$R(G; w, z) = \sum_{ij} m_{ij} w^i z^j, \tag{2.3}$$

we have $R(G; w, z) = w^{|G|} Q(G; w^{-1}, z)$, and $Q(G; t, z) = t^{|G|} R(G; t^{-1}, z)$. Using these relations, we see that (2.2) means exactly that

$$Q(G;t,z) = \begin{cases} Q(G-e;t,z) + Q(G/e;t,z) & \text{if } e \text{ is not a loop,} \\ (z+1)Q(G-e;t,z) & \text{if } e \text{ is a loop,} \end{cases} \tag{2.4}$$

and for $n = 1, 2, \ldots$ we also have

$$Q(E_n;t,z) = t^n, \tag{2.5}$$

where E_n is the empty graph of order n. It is immediate that the contraction-deletion formula (2.4) and the boundary conditions (2.5) determine Q. In 1947 Tutte [33] studied the dichromatic polynomial as an example of a *W-function*, a function W defined on $\mathscr{G}$, such that $W(G) = W(G-e) + W(G/e)$, whenever $e \in E(G)$ is not a loop.

As $m_{ij}(G) = q_{|G|-i,j}(G)$, Whitney's theorem given in (2.1) can be restated as follows:

$$p_G(\lambda) = \sum_{i,j}(-1)^{i+j}m_{ij}\lambda^{|G|-i} = \sum_{k,\ell}(-1)^{|G|+k+\ell}q_{k,\ell}\lambda^k = (-1)^{|G|}Q(G;-\lambda,-1). \tag{2.6}$$

This theorem of Whitney was extended by Tutte [33, 35] to a result concerning all vertex colourings of graphs, not only proper vertex colourings. Let $G = (V, E) \in \mathscr{G}$, $\lambda \in \mathbb{N}$, and set $F_\lambda = [\lambda]^V = \{1, 2, \ldots, \lambda\}^V$. Thus F_λ is the set of all vertex colourings of G with colours $1, 2, \ldots, \lambda$. For $f \in F_\lambda$ let $\phi(f)$ be the number of edges of G joining vertices of the same colour, so that f is a proper colouring if and only if $\phi(f) = 0$. Then

$$\sum_{f \in F_\lambda} x^{\phi(f)} = (x-1)^{|V|}Q(G; \frac{\lambda}{x-1}, x-1).$$

In a way reminiscent of Whitney's use of broken cycles, in 1954 Tutte [34] defined another polynomial whose definition depends *a priori* on an order we impose on the set of edges. Let $G = (V, E)$ be a connected graph (with loops and multiple edges). Let us endow E with a total order by choosing a $1-1$ map $\phi : E \to [|E|] = \{1, 2, \ldots, |E|\}$, and setting $x <_\phi y$, or simply $x < y$, if $\phi(x) < \phi(y)$. Given a spanning tree T of G (and our order given by ϕ), Tutte classified the edges of G as follows. An edge $e \in E(G)$ is *internally active* if it is the smallest edge between the two components of $T - e$. Also, an edge $f \in E(G) \backslash E(T)$ is *externally active* if it is the smallest edge in the unique cycle of $T \cup \{f\}$.

With these definitions, we are ready to introduce Tutte's *dichromate*, which is now always called the *Tutte polynomial* of G:

$$T(G;x,y) = \sum_{i,j} t_{ij}x^iy^j,$$

where t_{ij} is the number of spanning trees with precisely i internally active edges and precisely j externally active edges. Tutte proved the startling result that $T(G;x,y)$ is *independent of the order on E*.

It is natural to extend the domain of T to the entire set $\mathscr{G}$ by setting

$$T(G) = \prod_{i=1}^{k} T(G_i),$$

where $G_1, G_2, \ldots, G_k$ are the components of G.

Once we know that $T(G; x, y)$ is independent of the order, it is easy to show that T satisfies the contraction-deletion relations (1.1). The simplest case is when e is a loop. Then the spanning trees of G are exactly the spanning trees of $G - e$. As e itself forms a cycle, it is always externally active, contributing a factor of y. Also, deleting e does not affect the activity of any other edge f, as e can never lie in the cut or cycle determined by f. When e is a bridge the argument is similar, using the fact that the spanning trees of G are in bijection with those of G/e, where the bijection is given by contracting the edge e in each tree. Finally, when e is neither a bridge nor a loop, we make use of the order invariance, by taking e to be the last edge in the order ϕ. In this case e is never active. Also, the spanning trees of G containing e are in bijection with those of G/e, and the spanning trees of G not containing e are in bijection with those of $G - e$. Together with the fact that the presence of e cannot alter the activity of another edge f, as $f < e$, this shows that (1.1) is satisfied in this case as well.

Note that, when e is a bridge, we can replace the condition $T(G; x, y) = xT(G/e; x, y)$ by the condition $T(G; x, y) = xT(G - e; x, y)$. This is because the graphs G/e and $G - e$ have the same blocks, and the Tutte polynomial of a graph depends only on its blocks. We shall use the formulation with G/e, as it has the advantage of giving the Tutte polynomial of a connected graph in terms of the Tutte polynomials of smaller connected graphs.

It is trivial that $T(G; x, y)$ is determined by the recursion formula (1.1), together with the boundary condition (1.2) that $T(E_n; x, y) = 1$ for every n. As mentioned in the previous section, an alternative approach is to use only the relation (1.3) for ordinary edges, and the boundary condition (1.4). It is easy to see that these two approaches are equivalent. On the one hand, any solution to (1.1) and (1.2) clearly satisfies (1.3) and (1.4). On the other hand, suppose T' is a solution to (1.3) satisfying (1.4), so T' trivially satisfies (1.2), and let G be a graph with a bridge e. Then G and G/e have the same ordinary edges, as do the graphs obtained from them by applying the same sequence of contractions and deletions to each. We may thus apply the same sequence of reductions to G and G/e, using (1.3) to express $T'(G)$ as $\sum T'(B_i)$, for some sequence $(B_i)_1^k$ of basic graphs, and $T'(G/e)$ as $\sum T'(B_i/e)$, for the same sequence $(B_i)_1^k$. Since the boundary condition (1.4) ensures that $T(B_i) = xT(B_i/e)$, we thus have that $T'(G) = xT'(G/e)$. A similar argument shows that $T'(G) = yT'(G - e)$ when e is a loop, and thus that T' satisfies (1.1).

Tutte's dichromatic polynomial Q and dichromate (Tutte polynomial) T are related in a very simple way:

$$Q(G; t, z) = T(G; t + 1, z + 1)t^{k(G)}, \tag{2.7}$$

as can be seen from the recurrence relations obeyed by each, or from (1.7). In particular, (2.6) and (2.7) imply that

$$p_G(\lambda) = (-1)^{r(G)} \lambda^{k(G)} T(G; 1 - \lambda, 0). \tag{2.8}$$

The dichromatic polynomial and the dichromate are only two prominent members of a family of polynomials defined in various natural ways which can be obtained from each other by simple substitutions. In fact, taking the standard definitions of rank, nullity, deletion and contraction for *matroids*, the Tutte polynomial is easily extended to suitable classes of matroids. Going a little further, Brylawski and Oxley [4] define a

Tutte–Grothendieck invariant for a class $\mathcal{H}$ of matroids closed under isomorphism and taking minors, as a function f from $\mathcal{H}$ into a ring R such that, if $M \in \mathcal{H}$ and $e \in M$, then

$$f(M) = f(M - e) + f(M/e)$$

if e is neither a loop nor an isthmus (bridge), and

$$f(M) = f(M(e))f(M - e)$$

otherwise, where $M(e)$ is the submatroid on $\{e\}$. Brylawski and Oxley show that every Tutte–Grothendieck invariant is essentially an evaluation of the Tutte polynomial.

One frequently encounters polynomials that satisfy contraction-deletion formulae with coefficients other than 1: for example, the chromatic polynomial, or the polynomials f and f^* defined by Negami [21]. Since these polynomials are special cases of the polynomial U mentioned in the previous section, they are also simple transforms of the Tutte polynomial. Note that, in these cases, as for the Tutte polynomial, we could use $G - e$ instead of G/e when e is a bridge. For U we would get the same polynomial but with a change of variables, as the map U satisfies $U(G) = x\alpha^{-1}U(G - e)$ when e is a bridge.

Before proceeding further, let us show from first principles, without any reference to other polynomials or counting functions, that the contraction-deletion formula (1.1) and the boundary condition (1.2) do define a graph polynomial $T(G)$. The proof below is very simple, but as we spell it out in some detail, it is not that short. A similar argument could be used to prove the existence of $U(G)$.

Theorem 1. *There is a unique graph polynomial $T(G) = T(G; x, y)$ such that the contraction-deletion formula (1.1) holds and $T(E_n) = 1$ for every n.*

Proof. As the uniqueness of $T(G)$ is immediate, we have to show only its existence. We can view (1.1), together with the condition $T(E_n) = 1$, as a procedure for calculating $T(G)$: apply (1.1) to any edge $e \in E(G)$, then to any edge of each non-empty graph in the resulting expression, and so on, until $T(G)$ has been expressed in terms of x and y only. At first sight, it appears that for a fixed graph G this procedure may give many different results, depending on the edges chosen at each stage. In fact, we shall prove by induction on $e(G)$ that all these possible results are the same. Thus this procedure defines a graph polynomial $T(G)$, which satisfies the conditions of the theorem.

For $e(G) \leqslant 1$ the assertion is trivial, so assume that $e(G) \geqslant 2$, and that $T(G')$ is well defined for every graph G' with $e(G') < e(G)$. For $e \in E(G)$, let $T_e(G) = T_e(G; x, y)$ be given by (1.1); all we need to show is that $T_e(G)$ is independent of e, i.e., that

$$T_e(G) = T_f(G)$$

for any distinct edges $e, f \in E(G)$.

Let us write $T_{e,f}(G)$ for the expression obtained for $T(G)$ by first applying (1.1) to $e \in E(G)$ and then to $f \in E(G-e)$ or $f \in E(G/e)$. For example, if e is a bridge of G and f is a loop then $T_{e,f}(G) = xyT(G/e - f)$, if e and f are bridges then $T_{e,f}(G) = x^2T(G/e/f)$, if e and f are loops then $T_{e,f}(G) = y^2T(G - e - f)$, if e is a bridge and f is an ordinary edge of G/e (neither a bridge, nor a loop) then $T_{e,f}(G) = xT(G/e - f) + xT(G/e/f)$, and so on. Let $T_{f,e}(G)$ be defined similarly.

As, by our induction hypothesis, $T_e(G) = T_{e,f}(G)$ and $T_f(G) = T_{f,e}(G)$, it suffices to check that $T_{e,f}(G) = T_{f,e}(G)$. Now, this is clearly true if e and f are parallel (*i.e.*, have precisely the same end-vertices), or if the 'nature' of e is the same in G, G/f and $G - f$, and the 'nature' of f is the same in G, G/e and $G - e$. Here the 'nature' of an edge is whether it is a bridge, a loop or an ordinary edge. If e is a loop in G then it is also a loop in G/f and $G - f$, and if it is a bridge in G then it is also a bridge in G/f and $G - f$. Also, if e is an ordinary edge in G and f is not parallel to e, then e is an ordinary edge in G/f as well, and it is an ordinary edge in $G - f$ unless e is a bridge in $G - f$ (and so f is a bridge in $G - e$).

Hence, it suffices to check that $T_{e,f}(G) = T_{f,e}(G)$ in the case when G is obtained from a graph $G_1 \cup G_2$, with $V(G_1) \cap V(G_2) = \emptyset$, by adding to it edges e and f, both joining a component C_1 of G_1 to a component C_2 of G_2. Then

$$T_{e,f}(G) = T(G - e) + T(G/e) = xT(G - e/f) + T(G/e/f) + T(G/e - f),$$

and

$$T_{f,e}(G) = T(G - f) + T(G/f) = xT(G - f/e) + T(G/f/e) + T(G/f - e).$$

Since $G/e/f$ is the same graph as $G/f/e$, it suffices to show that

$$T(G/e - f) = T(G/f - e),$$

and

$$T(G - e/f) = T(G - f/e).$$

In fact, these two statements are equivalent, as the graphs involved are exactly the same. In particular, both $G/e - f = G - f/e$ and $G/f - e = G - e/f$ are of the form $G_1 \cup G_2$, with G_1 and G_2 sharing precisely one vertex (albeit different pairs of vertices in the two graphs). Now, the conditions on T imply easily that, if $H = H_1 \cup H_2$ with $|V(H_1) \cap V(H_2)| \leqslant 1$ and T is well defined on H, then $T(H) = T(H_1)T(H_2)$. From the induction hypothesis, T is well defined on the relevant graphs, so we have $T(G/e - f) = T(G_1)T(G_2) = T(G/f - e)$, completing the proof. $\square$

The Tutte polynomial extends not only the chromatic polynomial, as shown by (2.8), but also the flow polynomial of a graph. Given a graph G and an additively written finite abelian group A, an *A-flow* on G is a flow on G with values in A that satisfies Kirchhoff's current law at each vertex. Writing $q_G(A)$ for the number of nowhere-zero A-flows on G, it turns out that $q_G(A)$ depends only on the cardinality of A, so we may define $q_G(|A|) = q_G(A)$. The function $q_G(\lambda)$ is a polynomial in λ, called the *flow polynomial* of G. The flow polynomial is essentially the Tutte polynomial evaluated at $x = 0$:

$$q_G(\lambda) = (-1)^{n(G)} T(G; 0, 1 - \lambda).$$

As mentioned in the previous section, many specific evaluations of the Tutte polynomial have attractive interpretations in terms of various graph invariants. For example, if G is a connected graph, then $T(G;1,1)$ is the number of spanning trees, $T(G;2,1)$ is the number of forests, $T(G;1,2)$ is the number of connected spanning subgraphs, and $T(G;2,2)$ is the number of spanning subgraphs.

Although we shall not consider such questions here, we note that much work has been done concerning the complexity of evaluating the Tutte polynomial of a graph G. For example, Annan [2] considers the complexity of calculating the coefficients of $T(G;x,y)$, while Jaeger, Vertigan and Welsh [10] consider the complexity of evaluating $T(G;x,y)$ for given $x, y \in \mathbb{Q}$. In most cases, these calculations turn out to be #P-hard, but random algorithms for approximating the values have been described, by Alon, Frieze and Welsh [1], for example. For a survey of such results, see [37]. On a slightly different note, Sekine, Imai and Tani [27] give practical algorithms for calculating the Tutte polynomials of moderately sized graphs, including all graphs with at most 14 vertices and at most 91 edges.

It is difficult to overestimate the importance of the Tutte polynomial and its extensions, as they are fundamental in such diverse fields as graph theory, knot theory, percolation theory, coding theory and statistical mechanics. More often than not, one needs Tutte polynomials of graphs with additional structures and a variety of boundary conditions. As described in the Introduction, our main aim is to determine the most general forms of the Tutte polynomial under certain conditions.

Perhaps the two most important properties of the Tutte polynomial are the existence of a contraction-deletion formula and the existence of a spanning tree expansion for connected graphs. In addition, it also has an expansion in terms of the rank and nullity of its subgraphs. In generalizing the Tutte polynomial, one can make use of any of these properties. As the Tutte polynomial itself has a history of appearing in a variety of different guises, we shall investigate the relationships between these apparently different generalizations.

Given a set $\mathscr{G}_c$ of graphs with colours on the edges, what is the most general form of a map $f : \mathscr{G}_c \to R$, where R is a ring, such that f satisfies an appropriate contraction-deletion formula, and what is the most general form that has a spanning tree expansion for connected graphs? In the next section, we shall answer these questions for coloured graphs. In the subsequent sections, we shall study the connection with other generalizations of the Tutte polynomial, and with polynomials arising in knot theory.

3. A Tutte polynomial for coloured graphs

In this section we shall consider functions having a spanning tree expansion, or satisfying recurrence relations, with coefficients that are different for different edges e of a graph G. There are two possible approaches. The first is to take the coefficients as functions of the actual edge e, obtaining a graph function that is not an isomorphism invariant. The second approach, which seems more natural, is to consider graphs with colours on the edges, and to take the coefficients as functions of the colour $c(e)$ of the edge e. In this way we shall obtain functions of coloured graphs invariant under those graph isomorphisms which map each edge to an edge of the same colour.

Most of the time it will make little difference which approach we consider, and we shall phrase all our arguments in terms of the second approach. When there is a significant difference, we shall point this out.

3.1. Spanning tree expansions

For the rest of this paper, by a *coloured graph* we mean a graph $G = (V, E)$ together with a function $c : E \to \Lambda$, where Λ is a set, the *set of colours*. If $c(e) = \lambda \in \Lambda$ then the edge e has *colour* λ. We shall write $\mathcal{G}_c$ for the set of all such pairs (G, c).

As in Section 2, we endow E with a *total order* by taking a bijection ϕ from E to $\{1, \ldots |E|\}$. For $e, f \in E$, we write $e <_\phi f$, or simply $e < f$, if $\phi(e) < \phi(f)$. Let $\mathcal{G}_{c,o}$ be the set of all triples (G, c, ϕ), where (G, c) is a coloured graph, and ϕ is an order on $E(G)$.

In order to ensure that our graphs have spanning trees rather than only spanning forests, throughout this section we shall restrict our attention to the sets $\mathcal{G}_c^*$ and $\mathcal{G}_{c,o}^*$ of coloured graphs (with an order) whose underlying graph G is connected. Furthermore, for notational simplicity, we shall identify spanning subgraphs of G with their edge sets.

Let $T \subseteq E$ be a spanning tree of G. For each edge $e \in T$, by the *cut* of $T - e$ we mean the set of edges of G going between the two components of $T - e$, and we denote it by $\mathrm{cut}(T - e)$. Also, if $e \in E - T$ then we write $\mathrm{cyc}(T \cup e)$ for the unique cycle in $T \cup e$, and we call it the *cycle* of $T \cup e$.

We say that an edge $e \in T$ is *internally active* (with respect to T) if it is the first edge in $\mathrm{cut}(T - e)$, in the order ϕ. Otherwise, $e \in T$ is *internally inactive*. Also, $e \in E - T$ is *externally active* if e is the first edge in $\mathrm{cyc}(T \cup e)$, again in the order ϕ, and *externally inactive* otherwise.

In this section we shall use the notions of internal and external activity to define a coloured Tutte polynomial $W(G, c, \phi)$ in as general a way as possible. We shall then establish the least restrictive conditions under which this polynomial is independent of the order ϕ.

We start by defining the *weight* $w(G, c, \phi, T, e)$ of an edge e with respect to a spanning tree T as follows. If e has colour λ, then

$$
w(G, c, \phi, T, e) = \begin{cases} X_\lambda & \text{if } e \text{ is internally active,} \\ Y_\lambda & \text{if } e \text{ is externally active,} \\ x_\lambda & \text{if } e \text{ is internally inactive,} \\ y_\lambda & \text{if } e \text{ is externally inactive.} \end{cases}
$$

Initially we shall define a polynomial in the independent variables $\{X_\lambda, Y_\lambda, x_\lambda, y_\lambda : \lambda \in \Lambda\}$. Later we shall impose relations on these variables to ensure that the polynomial is independent of the order ϕ.

For a spanning tree T, we now define the *weight* of T as

$$
w(G, c, \phi, T) = \prod_{e \in E} w(G, c, \phi, T, e).
$$

Finally, the *coloured Tutte polynomial* $W_0(G, c, \phi) \in \mathbb{Z}[X_\lambda, Y_\lambda, x_\lambda, y_\lambda : \lambda \in \Lambda]$ is defined as the sum over all spanning trees T of $w(G, c, \phi, T)$. From now on we shall write $\mathbb{Z}_\Lambda$ for $\mathbb{Z}[X_\lambda, Y_\lambda, x_\lambda, y_\lambda : \lambda \in \Lambda]$.

Note that if Λ consists of one element or, equivalently, if we substitute

$$
\begin{aligned}
X_\lambda &= x, \\
Y_\lambda &= y, \\
x_\lambda &= 1, \\
y_\lambda &= 1, \tag{3.1}
\end{aligned}
$$

for all λ, then the weight of a tree T is just $x^i y^j$, where i and j are the numbers of internally and externally active edges, respectively. Thus, in this case, W_0 is just the usual Tutte polynomial $T(G; x, y)$ for G as an uncoloured graph.

Before we turn to the conditions for independence of the order, let us consider what happens if G is a *plane graph*, and we replace it by its dual G'. For the Tutte polynomial $T(G)$ we have $T(G; x, y) = T(G'; y, x)$, and this extends to the polynomial $W_0(G, c, \phi)$ we have just defined for coloured plane graphs. To ensure that $W_0(G', c, \phi)$ makes sense, we shall identify the edge set E' of G' with $E = E(G)$ in the obvious way. Now a spanning subgraph of G, that is, a set $F \subseteq E$, is connected if and only if its complement has no cycle when considered as a subgraph of G'. Dually, a subgraph of G is acyclic if and only if its complement is a connected subgraph of G'. Thus $T \subseteq E$ is a spanning tree in G if and only if $E - T$ is a spanning tree in G'. Also, since the cut $T - e$ in G and the cycle of $E - T \cup e$ in G' consist of the same edges, an edge e is active in G with respect to T if and only if it is active in G' with respect to $E - T$. Combining these observations with the definition of $W_0(G, c, \phi)$, we obtain

$$
W_0(G, c, \phi)(X_\lambda, Y_\lambda, x_\lambda, y_\lambda)_{\lambda \in \Lambda} = W_0(G', c, \phi)(Y_\lambda, X_\lambda, y_\lambda, x_\lambda)_{\lambda \in \Lambda}. \tag{3.2}
$$

We now turn to the dependence of the map $W_0 : \mathscr{G}_{c,o}^* \to \mathbb{Z}_\Lambda$ on the order. There is no reason why $W_0(G, c, \phi)$ should not depend on ϕ, and as the calculations in Section 3.3 will show, this is indeed the case. However, we know that some evaluations of W_0 do not depend on the order, for example (3.1). For a fixed (G, c), which evaluations of $W_0(G, c, \phi)$ are independent of ϕ will depend very much on the structure of (G, c). For example, if (G, c) is monochromatic, then any evaluation will do. (This can be deduced from Tutte's result, and also follows from Theorem 2 below.) Also, if G has a bridge e of colour λ, then since any spanning tree T uses e, and e is the only edge in $\mathrm{cut}(T - e)$, and hence is active, any evaluation with $X_\lambda = 0$ will make $W_0(G, c, \phi)$ zero for any order. We would like to know for which evaluations $W_0(G, c, \phi)$ is independent of ϕ for *all* coloured graphs. The most general form of this question is: for which ring homomorphisms $f : \mathbb{Z}_\Lambda \to R$, where R is any ring, is the map $f \circ W_0$ independent of ϕ? This is the same question as the following: for which ideals $I \subseteq \mathbb{Z}_\Lambda$ is the induced map $\overline{W}_0 : \mathscr{G}_{c,o}^* \to \mathbb{Z}_\Lambda / I$ independent of the order? We now state our first main result, which answers this question precisely, and allows us to describe the most general connected graph invariant with a spanning tree expansion.

Theorem 2. *Let $I \subseteq \mathbb{Z}_\Lambda$ be an ideal, and let $\overline{W}_0 : \mathscr{G}_{c,o}^* \to \mathbb{Z}_\Lambda / I$ be the composition*

$$
\mathscr{G}_{c,o}^* \xrightarrow{W_0} \mathbb{Z}_\Lambda \xrightarrow{q} \mathbb{Z}_\Lambda / I,
$$

where q is the quotient map. Then $\overline{W}_0(G, c, \phi) = \overline{W}_0(G, c, \phi')$ *holds for all connected coloured graphs* (G, c), *and all pairs of orders* ϕ, ϕ' *on* $E(G)$, *if and only if*

$$X_\lambda y_\mu - y_\lambda X_\mu - x_\lambda Y_\mu + Y_\lambda x_\mu \in I, \tag{3.3}$$

$$Y_\nu(x_\lambda Y_\mu - Y_\lambda x_\mu - x_\lambda y_\mu + y_\lambda x_\mu) \in I, \tag{3.4}$$

and

$$X_\nu(x_\lambda Y_\mu - Y_\lambda x_\mu - x_\lambda y_\mu + y_\lambda x_\mu) \in I, \tag{3.5}$$

for all λ, μ, $\nu \in \Lambda$.

Note that if we consider uncoloured graphs, and coefficients $X_e, \ldots$ depending on the edge e, rather than on its colour, we obtain a result exactly like Theorem 2, but with λ, μ, ν replaced by all triples e, f, g of *distinct* edges.

Returning to coloured graphs, from now on we shall write I_0 for the minimal ideal $I \subseteq \mathbb{Z}_\Lambda$ satisfying (3.3)–(3.5). Also, given an ideal $I \supseteq I_0$, so that $\overline{W}_0(G, c, \phi) \in \mathbb{Z}_\Lambda/I$ does not depend on ϕ, we shall write W for the map $W : \mathscr{G}_c^* \to \mathbb{Z}_\Lambda/I$ defined by $W(G, c) = \overline{W}_0(G, c, \phi)$ for any order ϕ. If the ideal I is not specified, then we take $I = I_0$. The content of Theorem 2 is that the map $W : \mathscr{G}_c^* \to \mathbb{Z}_\Lambda/I_0$ obtained in this case is the most general connected graph invariant with a spanning tree expansion of the form described in Section 3.1, with arbitrary coefficients. Indeed, any such invariant W' is, by definition, of the form $W' = f \circ W_0$, for some ring homomorphism f. Theorem 2 then tells us that the kernel of f contains I_0, so W' can be written in the form $f' \circ W$. We have thus answered the first question raised in the Introduction, finding an invariant with a spanning tree expansion from which all others may be obtained.

In the case where $\mathbb{Z}_\Lambda/I$ is an integral domain, the conditions above simplify, since from (3.4) and (3.5) we can conclude that either $x_\lambda Y_\mu - Y_\lambda x_\mu - x_\lambda y_\mu + y_\lambda x_\mu \in I$ for all λ and μ, or X_λ and Y_λ are in I for all λ. Rephrasing this slightly, we have the following immediate consequence of Theorem 2.

Corollary 3. *Let* $I \subseteq \mathbb{Z}_\Lambda$ *be any ideal, and let* $\overline{W}_0$ *be as before. Then* $\overline{W}_0(G, c, \phi) = \overline{W}_0(G, c, \phi')$ *holds for all connected coloured graphs* (G, c), *and all pairs of orders* ϕ, ϕ' *on* $E(G)$, *provided that either*

$$X_\lambda y_\mu - y_\lambda X_\mu = x_\lambda Y_\mu - Y_\lambda x_\mu = x_\lambda y_\mu - y_\lambda x_\mu \tag{3.6}$$

holds in $\mathbb{Z}_\Lambda/I$ *for all colours* λ *and* μ, *or*

$$X_\lambda = Y_\lambda = 0 \tag{3.7}$$

holds in $\mathbb{Z}_\Lambda/I$ *for all colours* λ.

Furthermore, if $\mathbb{Z}_\Lambda/I$ *is an integral domain, then these conditions are necessary.* □

Note that for the formulation with uncoloured graphs and coefficients $X_e, \ldots$ depending on the edge e, the equivalent of Corollary 3 does not hold. This is because, in this case, we cannot take ν equal to λ or μ in condition (3.4) or (3.5).

The force of Corollary 3 is shown by the following argument. Given any non-empty coloured graph (G, c), and any order ϕ on $E = E(G)$, there is some edge e with $\phi(e) = 1$. Suppose this edge has colour λ. Then e will be active (internally or externally) with respect to any spanning tree of G, and will contribute a factor of either X_λ or Y_λ to its weight. Thus if (3.7) holds, then $\overline{W}_0(G, c, \phi)$ will be zero in $\mathbb{Z}_\Lambda / I$ whenever G is non-empty, for any order ϕ. Corollary 3 thus says essentially that, in the integral domain case, condition (3.6) is necessary and sufficient for $\overline{W}_0$ to be order-independent.

We shall prove Theorem 2 in two sections, devoting the next section to the proof of the sufficiency of the conditions given for order independence, and the subsequent section to the proof of their necessity.

3.2. The sufficiency of the conditions for order independence

In order to prove that the conditions given in Theorem 2 are sufficient to ensure that $\overline{W}_0(G, c, \phi)$ is independent of the order ϕ, we shall follow the proof that the spanning tree expansion of the Tutte polynomial for uncoloured graphs is independent of the order.

Let ϕ and ϕ' be two orders on E obtained from each other by a *transposition*, so that they agree except on two edges, e and f, say, on which we have $\phi(e) = i$, $\phi(f) = i + 1$, and $\phi'(e) = i + 1$, $\phi'(f) = i$. Since any two orders can be obtained from each other by applying a sequence of transpositions, it suffices to find conditions ensuring that $W_0(G, c, \phi) - W_0(G, c, \phi') \in I$ for all connected coloured graphs (G, c) and pairs of orders $\{\phi, \phi'\}$ related by a transposition.

Now an edge $e_1 \in T$ is internally active if and only if there is no other edge $e_2 \in \text{cut}(T - e_1)$ preceding e_1. Similarly, an edge $e_2 \notin T$ is externally active if and only if there is no other edge $e_1 \in \text{cyc}(T \cup e_2)$ preceding e_2. We say that a pair of edges $\{e_1, e_2\}$ is *related* (with respect to T) if $e_1 \in T$, $e_2 \notin T$, and e_2 is in $\text{cut}(T - e_1)$. Since this is the same condition as e_1 being in $\text{cyc}(T \cup e_2)$, an edge is active if and only if it is not related to any edge preceding it. Hence the activities of all the edges of G are determined by comparing related pairs of edges.

The only edges whose orders change when we switch from ϕ to ϕ' are e and f; furthermore, the only *pair* of edges whose relative order changes is the pair $\{e, f\}$. Thus, for any tree T with respect to which e and f are not related, we have $w(G, c, \phi, T) = w(G, c, \phi', T)$. Hence it suffices to consider trees T for which e and f are related. Note that all such trees contain precisely one of e and f. Furthermore, there is a bijection from those containing e to those containing f, given by removing e and adding f; the fact that e and f are related ensures that the result is a spanning tree in which e and f are still related. Thus, to show that $W_0(G, c, \phi) - W_0(G, c, \phi') \in I$, it suffices to prove that, for pairs $\{T, T'\}$ where T is a spanning tree containing e and not f, with $e \in \text{cyc}(T \cup f)$, and T' is the spanning tree $T - e \cup f$, we have

$$w(G, c, \phi, T) + w(G, c, \phi, T') - w(G, c, \phi', T) - w(G, c, \phi', T') \in I. \tag{3.8}$$

We would like to be able to ignore the weights of the other edges, and concentrate on e and f. It is true that the activity of an edge g different from e and f is not affected by switching from the order ϕ to the order ϕ', since comparing $\phi(g) \neq i, i + 1$ with i or $i + 1$ gives the same result. However, such an edge g may be active in one of T, T' and inactive in the other. Suppose then that $g \in T$ is an edge whose activity changes. Now g is preceded by an edge in precisely one of $\mathrm{cut}(T - g)$ and $\mathrm{cut}(T' - g)$. For these cuts to be different, g must belong to $\mathrm{cyc}(T \cup f)$. Since each cut then contains one of e and f, and g is the first edge in one of the cuts, we have $\phi(g) < i$. Also, the symmetric difference between these cuts is just $\mathrm{cut}(T - e)$, so some edge h in this cut satisfies $\phi(h) < \phi(g) < i$. Thus there are edges of order less than i both in $\mathrm{cyc}(T \cup f)$ (the edge g is such) and in $\mathrm{cut}(T - e)$ (the edge h is such). A similar argument holds if an edge $g \notin T$ changes activity between T and T'.

We now consider four cases, setting $\lambda = c(e)$, $\mu = c(f)$ and writing $\phi_0(F)$ for the minimal order of an element of $F \subseteq E$.

Case 1. Both $\phi_0(\mathrm{cyc}(T \cup f)) < i$ and $\phi_0(\mathrm{cut}(T - e)) < i$. (The remarks above show that if some edge other than e, f changes activity between T and T', then this case holds.) In this case e and f are inactive with respect to T, whether we consider the order ϕ or ϕ'. They are also inactive with respect to T', as $\mathrm{cyc}(T' \cup e) = \mathrm{cyc}(T \cup f)$, and $\mathrm{cut}(T' - f) = \mathrm{cut}(T - e)$. Now, since no other edge changes activity when we switch from ϕ to ϕ', we have

$$w(G, c, \phi, T) = w(G, c, \phi', T), \text{ and } w(G, c, \phi, T') = w(G, c, \phi', T'),$$

from which (3.8) follows.

For the remaining cases, we can simplify (3.8) by factoring out the product w_0 of the weights of the edges other than e and f, since, as shown above, these weights cannot change.

Case 2. We have $\phi_0(\mathrm{cut}(T - e)) < i$, but $\phi_0(\mathrm{cyc}(T \cup f)) \geqslant i$. Now f is the first edge in $\mathrm{cyc}(T \cup f)$ in the order ϕ', but not in the order ϕ. Similarly, e is the first edge in $\mathrm{cyc}(T' \cup e) = \mathrm{cyc}(T \cup f)$ only in ϕ. Thus (3.8) reduces to

$$w_0(x_\lambda y_\mu + Y_\lambda x_\mu - x_\lambda Y_\mu - y_\lambda x_\mu) \in I,$$

or

$$w_0(x_\lambda Y_\mu - Y_\lambda x_\mu - x_\lambda y_\mu + y_\lambda x_\mu) \in I. \tag{3.9}$$

Now, as $\phi_0(\mathrm{cut}(T - e)) < i$, the edge h with $\phi(h) = 1$ is distinct from e and f. Let $c(h) = v$. As h is active, whether it lies in T or not, it contributes a factor of either X_v or Y_v to w_0. Thus (3.9) follows from (3.4) and (3.5).

Case 3. We have $\phi_0(\mathrm{cyc}(T \cup f)) < i$, but $\phi_0(\mathrm{cut}(T - e)) \geqslant i$. This time e or f is internally active whenever it has order i, so (3.8) becomes

$$w_0(X_\lambda y_\mu + y_\lambda x_\mu - x_\lambda y_\mu - y_\lambda X_\mu) \in I. \tag{3.10}$$

As before, w_0 has a factor of X_v or Y_v for some v. Also note that, given (3.3), equation (3.10) is equivalent to (3.9). Thus (3.10) follows from (3.3), (3.4) and (3.5).

Case 4. Both $\phi_0(\text{cyc}(T \cup f)) \geqslant i$ and $\phi_0(\text{cut}(T - e)) \geqslant i$. This time (3.8) becomes

$$X_\lambda y_\mu + Y_\lambda x_\mu - x_\lambda Y_\mu - y_\lambda X_\mu \in I, \qquad (3.11)$$

which follows from (3.3).

We have now shown that, provided (3.3), (3.4) and (3.5) hold for all colours λ, μ, ν, we have $W_0(G, c, \phi) - W_0(G, c, \phi') \in I$ for all pairs of orders $\{\phi, \phi'\}$ related by a transposition, and hence for all pairs of orders. This proves the sufficiency part of Theorem 2. $\qquad\square$

3.3. The necessity of the conditions for order independence
In this section we complete the proof of Theorem 2, by showing that the conditions given are necessary to ensure the order independence of $\overline{W}_0(G, c, \phi)$, by considering some specific small coloured graphs.

Example 1. We first take G to be a *double edge*, that is, a graph with two vertices v_1 and v_2, and two edges, e_1 and e_2, which both go from v_1 to v_2. Let e_1 have colour λ and e_2 colour μ.

Taking $\phi(e_1) = 1$, $\phi(e_2) = 2$ and $\phi'(e_1) = 2$, $\phi'(e_2) = 1$, we have

$$W_0(G, c, \phi) = X_\lambda y_\mu + Y_\lambda x_\mu$$

and

$$W_0(G, c, \phi') = x_\lambda Y_\mu + y_\lambda X_\mu.$$

We thus require

$$X_\lambda y_\mu + Y_\lambda x_\mu - x_\lambda Y_\mu - y_\lambda X_\mu \in I,$$

which is exactly (3.11). Hence we require

$$X_\lambda y_\mu - y_\lambda X_\mu - x_\lambda Y_\mu + Y_\lambda x_\mu \in I,$$

for all λ and μ, which is just (3.3).

Example 2. Next we consider a triple edge, where $c(e_1) = \nu$, $c(e_2) = \lambda$, and $c(e_3) = \mu$. For $\phi(e_1) = 1$, $\phi(e_2) = 2$, $\phi(e_3) = 3$ and $\phi'(e_1) = 1$, $\phi'(e_2) = 3$, $\phi'(e_3) = 2$, we obtain

$$W_0(G, c, \phi) = X_\nu y_\lambda y_\mu + Y_\nu x_\lambda y_\mu + Y_\nu Y_\lambda x_\mu,$$

and

$$W_0(G, c, \phi') = X_\nu y_\lambda y_\mu + Y_\nu x_\lambda Y_\mu + Y_\nu y_\lambda x_\mu.$$

Hence we require

$$Y_\nu x_\lambda y_\mu + Y_\nu Y_\lambda x_\mu - Y_\nu x_\lambda Y_\mu - Y_\nu y_\lambda x_\mu \in I,$$

i.e., that (3.4) holds:

$$Y_\nu(x_\lambda Y_\mu - Y_\lambda x_\mu - x_\lambda y_\mu + y_\lambda x_\mu) \in I.$$

Now, considering the dual graph to the triple edge, using the relation (3.2), a similar argument shows that we must also have

$$X_\nu(y_\lambda X_\mu - X_\lambda y_\mu - y_\lambda x_\mu + x_\lambda y_\mu) \in I.$$

Given (3.3), which we have already shown to be necessary, this is equivalent to (3.5). Thus we have shown that for the map $\overline{W}_0 : \mathscr{G}^*_{c,o} \to \mathbb{Z}_\Lambda/I$ to be independent of the order ϕ, we must have (3.3), (3.4) and (3.5). This completes the proof of Theorem 2. $\qquad\square$

In the next section we shall extend W to the whole set $\mathscr{G}_c$, allowing disconnected graphs.

3.4. Disconnected graphs

So far we have been considering only connected graphs, so that the concept of spanning trees made sense. In order to extend W to disconnected graphs, we consider *spanning forests* instead, where a spanning forest $F \subseteq G$ is the union of a spanning tree for each component of G, so F is a forest with $k(F) = k(G)$. Given $(G, c, \phi) \in \mathscr{G}_{c,o}$, and a spanning forest $F \subseteq G$, we can define the activity of an edge $e \in E - F$ as before. This time, however, an edge $e \in F$ is active if it is the first edge in $\mathrm{cut}(F - e)$, which we define as the set of all edges between the two components of $F - e$ joined by e. We define the weight $w(G, c, \phi, F, e)$ as before, and the weight of F as

$$w(G, c, \phi, F) = \alpha_{k(F)} \prod_{e \in E} w(G, c, \phi, F, e).$$

We now define a map $W'_0 : \mathscr{G}_{c,o} \to \mathbb{Z}_{\Lambda, \alpha_i} = \mathbb{Z}_\Lambda[\alpha_i : i = 1, 2, \ldots]$, setting $W'_0(G, c, \phi)$ to be the sum of the weights of the spanning forests of F.

Suppose $(G, c, \phi) \in \mathscr{G}_{c,o}$, and that G has k components, $G_1, \ldots, G_k$. Then a spanning tree F of G is just the union of spanning trees $T_i \subseteq G_i$. Also, the activity of an edge $e \in G_i$ with respect to F is just its activity with respect to T_i, using the induced order on $E(G_i)$. Thus we have the following relationship between W'_0 and W_0:

$$W'_0(G, c, \phi) = \alpha_{k(G)} \prod_{i=1}^{k} W_0(G_i, c, \phi_i), \tag{3.12}$$

where ϕ_i is the order induced on $E(G_i)$ by ϕ. Thus W'_0 is essentially given by W_0, except that we can now choose a different normalization for graphs with different numbers of components. In some contexts restricting this normalization to be of the form $\alpha_n = C\alpha^{n-1}$ will give a polynomial with nicer properties. Since this does not destroy any information about the graph, we shall do this when convenient. For the moment, we leave W'_0 in its most general form. From now on, we shall also write W_0 for the extended map W'_0, even though there is a slight inconsistency with the normalization. Whenever it matters, it will always be clear which normalization we are considering.

As a consequence of the fact that it is only the orders induced on the components that matter when evaluating W_0, we can immediately extend Theorem 2 to graphs with more than one component to obtain the following result.

Corollary 4. *Let $I \subseteq \mathbb{Z}_{\Lambda, \alpha_i}$ be an ideal, and let $\overline{W}_0 : \mathscr{G}_{c,o} \to \mathbb{Z}_{\Lambda, \alpha_i}/I$ be the composition*

$$\mathscr{G}_{c,o} \xrightarrow{W_0} \mathbb{Z}_{\Lambda, \alpha_i} \xrightarrow{q} \mathbb{Z}_{\Lambda, \alpha_i}/I,$$

where q is the quotient map. Then $\overline{W}_0(G, c, \phi) = \overline{W}_0(G, c, \phi')$ holds for all coloured graphs

(G, c), and all pairs of orders ϕ, ϕ' on $E(G)$, if and only if $I_0' \subseteq I$, where I_0' is the ideal of $\mathbb{Z}_{\Lambda, \alpha_i}$ generated by $\bigcup_{i=1}^{\infty} \alpha_i I_0 \subset \mathbb{Z}_{\Lambda, \alpha_i}$.

Proof. For sufficiency, consider (G, c) with k components, and two orders ϕ, ϕ' differing by a transposition. If the edges whose order changes lie in different components, then we have $W_0(G, c, \phi) = W_0(G, c, \phi')$. If they lie in the same component, then the result follows from (3.12) and Theorem 2.

For necessity we consider the same graphs as in Section 3.3. With (3.12) and Theorem 2, we see that for $\overline{W}_0$ not to depend on the order on these graphs, we must have $\alpha_1 I_0 \subseteq I$. Considering the same graphs with isolated vertices added, we deduce that $\alpha_i I_0 \subseteq I$, completing the proof. $\qquad \square$

As in the connected case, given an ideal $I \subseteq \mathbb{Z}_{\Lambda, \alpha_i}$ containing I_0', so that $\overline{W}_0(G, c, \phi)$ does not depend on ϕ, we shall write W for the map $W : \mathscr{G}_c \to \mathbb{Z}_{\Lambda, \alpha_i} / I$ defined by $W(G, c) = \overline{W}_0(G, c, \phi)$ for any order ϕ. Again, when I is not specified, we take $I = I_0'$. Note that in this case, the map W we obtain is the most general coloured graph invariant with a spanning forest expansion of the form described above, allowing arbitrary normalization on the empty graphs E_n.

We now turn to the second definition of the Tutte polynomial given in the Introduction, *i.e.*, that via recurrence relations.

3.5. Recurrence relations

In this section we have two apparently distinct aims. On the one hand, we would like to investigate the properties of the invariant $W(G, c)$ defined above and, in particular, to show that this invariant satisfies certain recurrence relations, analogous to the relations (1.1) obeyed by the Tutte polynomial. On the other hand, we would like to answer the second question raised in the Introduction, by finding the most general coloured graph invariant satisfying such relations. It turns out that these aims coincide: with the aid of Corollary 4, we shall show that $W(G, c)$ is exactly the invariant we are looking for.

To state recurrence relations for coloured graphs, given $(G, c) \in \mathscr{G}_c$, and an edge $e \in E(G)$, we consider G/e and $G - e$ to be coloured with the restrictions of c to $E(G/e)$ and $E(G - e)$ (which, as noted earlier, are naturally identified). The basic idea will be to use the relations (stated precisely below) to define a polynomial ω, and then show that ω is equal to W. The problem is that ω need not exist, since $\omega(G, c)$ must obey several relations defining it in terms of the values of ω on graphs (G', c') with fewer edges. We can get round this by re-introducing an order ϕ, and using this to choose one relation for each $(G, c, \phi) \in \mathscr{G}_{c,o}$.

Lemma 5. *For any ideal $I \subseteq \mathbb{Z}_{\Lambda, \alpha_i}$, the map $\overline{W}_0 : \mathscr{G}_{c,o} \to \mathbb{Z}_{\Lambda, \alpha_i} / I$ is the unique map $\omega_0 : \mathscr{G}_{c,o} \to \mathbb{Z}_{\Lambda, \alpha_i} / I$ satisfying the equations*

$$\omega_0(E_n, c, \phi) = \alpha_n$$

and

$$\omega_0(G,c,\phi) = \begin{cases} X_\lambda \omega_0(G/e,c,\phi) & \text{if } e \text{ is a bridge,} \\ Y_\lambda \omega_0(G-e,c,\phi) & \text{if } e \text{ is a loop,} \\ x_\lambda \omega_0(G/e,c,\phi) + y_\lambda \omega_0(G-e,c,\phi) & \text{if } e \text{ is neither,} \end{cases} \quad (3.13)$$

where (G,c) *is any non-empty coloured graph,* ϕ *is any order on* $E(G)$, *e is the last edge in the order* ϕ, *and e has colour* λ.

Proof. It is clear that the map ω_0, if it exists, is unique. (In fact, the relations above can be considered as an inductive definition of ω_0, so ω_0 will always exist.) Since we have that $\overline{W}_0(E_n,c,\phi) = \alpha_n$, it only remains to check that $\overline{W}_0$ satisfies (3.13). Since these relations only modify one component of G at a time, equation (3.12) shows that it is sufficient to check that $\overline{W}_0$ satisfies (3.13) for connected graphs.

Let $(G,c,\phi) \in \mathcal{G}^*_{c,o}$, e, and λ be as above. Suppose first that e is a bridge. Then every spanning tree T of G contains e, and the spanning trees of G are in one to one correspondence with the spanning trees of G/e, by removing the edge e. Also, as e is always the only edge in $\text{cut}(T-e)$, it is always active. Furthermore, the cuts $T-f$ in G and $T-e-f$ in G/e are exactly the same, as are the cycles of $T \cup g$ in G and $T-e \cup g$ in G'. Thus the edges other than e have the same activity with respect to T in G as with respect to $T-e$ in G/e. This shows that $w(G,c,\phi,T) = X_\lambda w(G/e,c,\phi,T-e)$. Summing over trees, we see that $W_0(G,c,\phi) = X_\lambda W_0(G/e,c,\phi)$, from which we deduce that $\overline{W}_0$ satisfies (3.13) in the case that e is a bridge.

In the case when e is a loop the argument is similar. This time no tree contains e, and e is the only edge in $\text{cyc}(T \cup e)$, and so it is always externally active, and contributes a factor of Y_λ.

In proving the third relation, we use the fact that e is the last edge in the order ϕ, and hence is never active. Now the spanning trees T containing e are in bijection with the spanning trees of G/e. Also, for such a tree T and another edge $f \in T$, the cuts $T-f$ in G and $T-e-f$ in G/e consist of exactly the same edges. For $f \notin T$, the cycles in $T \cup f \subseteq G$ and in $T-e \cup f \subseteq G/e$ agree, or differ only in that the first contains the edge e. Since e comes last in the order, f has the same activity in either case. Similarly, the spanning trees not containing e are in bijection with the spanning trees of $G-e$, with the activities of other edges remaining the same. Combining these observations gives the third case above. $\square$

We have now done all the work necessary to prove the following result, concerning the relations obeyed by $W(G,c)$ when it is well defined.

Theorem 6. *Let $I \subseteq \mathbb{Z}_{\Lambda,\alpha_i}$ be any ideal. If $I'_0 \subseteq I$, then the map $W : \mathcal{G}_c \to \mathbb{Z}_{\Lambda,\alpha_i}/I$ is the unique map $\omega : \mathcal{G}_c \to \mathbb{Z}_{\Lambda,\alpha_i}/I$ satisfying the equations*

$$\omega(E_n,c) = \alpha_n$$

and

$$\omega(G,c) = \begin{cases} X_\lambda \omega(G/e,c) & \text{if } e \text{ is a bridge,} \\ Y_\lambda \omega(G-e,c) & \text{if } e \text{ is a loop,} \\ x_\lambda \omega(G/e,c) + y_\lambda \omega(G-e,c) & \text{if } e \text{ is neither,} \end{cases} \qquad (3.14)$$

for all non-empty coloured graphs (G,c) and edges $e \in E(G)$ with $c(e) = \lambda$. Furthermore, if $I'_0 \nsubseteq I$, then there is no such map ω.

Proof. Clearly the solution ω is unique, if it exists. If $I'_0 \subseteq I$ then, from Corollary 4, $\overline{W}_0(G,c,\phi)$ does not depend on ϕ. Combined with Lemma 5, this implies that $W(G,c)$ solves (3.14). Indeed, when we consider a particular (G,c) and $e \in E(G)$, we pick an order ϕ in which e comes last, and use the fact that $\overline{W}_0(G,c,\phi)$ solves (3.13).

Conversely, suppose that $\omega'(G,c)$ is a solution of (3.14), and define $\omega'_0(G,c,\phi) = \omega'(G,c)$ for all orders ϕ. Then ω'_0 is a solution of (3.13), so $\omega'_0 = \overline{W}_0$. However, by definition $\omega'_0(G,c,\phi)$ does not depend on ϕ, so in this case $\overline{W}_0(G,c,\phi)$ does not depend on ϕ. Hence, by Corollary 4, we must have $I'_0 \subseteq I$. This completes the proof of Theorem 6. $\qquad \square$

The main force of this result is that the map $W : \mathcal{G}_c \longrightarrow \mathbb{Z}_{\Lambda,\alpha_i}/I'_0$ is the most general coloured graph invariant satisfying recurrence relations of the above form. Indeed, if W' is another such map, we may write X_λ, Y_λ, etc. for the coefficients in the relations obeyed by W'. We may then apply Theorem 6, which makes no assumptions about the relationship between X_λ, Y_λ, etc., to deduce that W' can be obtained from W by composing with a ring homomorphism.

Remark 1. Theorem 6 generalizes Theorem 1, by showing that the relations (3.14), which generalize (1.1), have a solution under certain conditions. This result could be proved directly, by adapting the proof of Theorem 1 given in the Introduction. However, this proof would be significantly longer, due to cases like that of two parallel edges of different colours, which corresponds to a trivial case in the proof of Theorem 1.

Remark 2. In the proof of the necessity part of Theorem 2 we only used the fact that W was defined on graphs with up to three edges. We thus have that Theorem 6 holds not only for invariants defined on all coloured graphs, but for invariants defined on any class of coloured graphs closed under contraction and deletion, and containing all coloured graphs with up to three edges. This will be important in Section 6, where we consider the set of signed planar graphs.

Remark 3. All our results for connected graphs can be extended to matroids. To see this, note that W is defined in terms of spanning trees, cuts and cycles. All these concepts can be expressed in terms of which sets of edges are *dependent*, that is, contain a cycle, and which are *independent*. Thus, like the usual Tutte polynomial, $W(G,c)$ can be calculated in terms of the cycle matroid of (G,c). Furthermore, all the properties of independent sets of edges we use hold for arbitrary matroids. We can thus extend W to arbitrary coloured matroids, obtaining results like Theorem 2 and Theorem 6, with all the α_i replaced by 1, and I_0' by I_0. We have thus found the universal coloured matroid invariant W with a 'spanning tree' (*i.e.*, basis) expansion as described in Section 3.1. We have also shown that W is the universal coloured matroid invariant satisfying recurrence relations of the form (3.14).

Remark 4. Zaslavsky [40] classifies all coloured matroid invariants satisfying recurrence relations of a form equivalent to (3.14), but with coefficients in a field K. This appears to be very close to the question we have answered. Indeed, since all fields are rings, Theorem 6 implies that all such invariants can be obtained from W. The reverse need not be the case, however. In fact, whether we take fields or rings seems to make rather a significant difference. For the uncoloured case (coefficients $X_e, \ldots$ depending on e), Zaslavsky obtains *seven* different classes of invariant, and in the coloured case, four. In contrast, we obtain a single universal invariant W in each case. Zaslavsky's results correspond to solving (3.3)–(3.5) in a field K, first with λ, μ, ν distinct, and then with λ, μ, ν arbitrary.

Remark 5. Schwärzler and Welsh [26] also consider coloured matroid invariants satisfying recurrence relations, in the special case $\Lambda = \{+, -\}$. Although their relations have a slightly different form, their Proposition 2.1 is equivalent to the assertion that W gives a well-defined signed matroid invariant with values in $\mathbb{Z}_\Lambda / I$ if and only if $I \supseteq I'$, the ideal generated by r and s, with

$$r = X_- y_+ - y_- X_+ - x_- Y_+ + Y_- x_+$$

and

$$s = X_- y_+ - y_- X_+ - x_- y_+ + y_- x_+.$$

Note that this contradicts Theorem 6, since $I' \neq I_0$, the ideal generated by the set $\{r, X_+ s, X_- s, Y_+ s, Y_- s\}$. To see that these ideals are different, note that substituting $X_+ = X_- = Y_+ = Y_- = 0$ maps every element of I_0 to 0, but not the element s of I'. What Schwärzler and Welsh actually prove is that W is well defined when $I \supseteq I'$, and that if it is well defined, then $I \supseteq I''$, where I'' is generated by r and $X_+ s$. Since $I'' \subset I_0 \subset I'$, this is consistent with Theorem 6, as it should be!

Remark 6. We already know that $W(G, c)$ is given by its values on the components of G. The relations above allow us to go slightly further. Since whether an edge $e \in E$ is a loop or a bridge or neither can be determined by looking only at the block of G containing e, we can apply Theorem 6 in each block in turn, to deduce that $W(G, c)$ is given by

$$W(G, c) = \alpha_{k(G)} \prod_{i=1}^{b} \frac{W(B_i, c)}{\alpha_1}, \tag{3.15}$$

where $B_1, \ldots B_b$ are the blocks of G.

Note that the freedom to specify the values of W on each empty graph independently is not surprising, since the relations (3.14) only involve graphs with the same number of components.

In this section we have shown that the recurrence relations (3.14) are essentially equivalent to the spanning tree expansion as a definition of a coloured graph polynomial. We have then used Corollary 4 of Theorem 2 to give necessary and sufficient conditions for these relations to have a solution.

3.6. Sample calculations

In this section we shall use Theorem 6 to calculate W on some slightly more complicated graphs than those considered in Section 3.3. For simplicity, and because the results will be useful later, we take the set of colours Λ to have just two elements, $+$ and $-$. Also, as the examples we consider will all be connected, their polynomials will all have a factor of α_1, which we set to 1 for simplicity.

Example 3. Any tree $T_{r,s}$ with r positive edges and s negative edges. In this case we can calculate W directly from the definition; there is only one spanning tree, and every edge is internally active, so

$$W(T_{r,s}) = X_+^r X_-^s. \tag{3.16}$$

In particular, for a path $P_{r,s}$ with r positive edges and s negative edges, we have the same expression for W. Note that (3.16) also follows immediately from (3.15) above.

Example 4. A cycle $C_{r,0}$, consisting of r positive edges. For $r = 1$, we take $C_{1,0}$ to be just a positive loop, and we have $W(C_{1,0}) = Y_+$. For $r \geqslant 1$, by Theorem 6,

$$
\begin{aligned}
W(C_{r,0}) &= x_+ W(C_{r-1,0}) + y_+ W(P_{r-1,0}) \\
&= x_+ W(C_{r-1,0}) + y_+ X_+^{r-1}.
\end{aligned}
$$

Hence, by induction,

$$W(C_{r,0}) = \sum_{i=0}^{r-2} x_+^i y_+ X_+^{r-i-1} + x_+^{r-1} Y_+, \tag{3.17}$$

for all $r \geqslant 1$. Similarly, for a cycle $C_{0,s}$ consisting of s negative edges, we have

$$W(C_{0,s}) = \sum_{i=0}^{s-2} x_-^i y_- X_-^{s-i-1} + x_-^{s-1} Y_-.$$

Example 5. A cycle $C_{r,s}$ with r positive edges and s negative ones. We have just calculated the case $s = 0$ above. If $s \geqslant 1$, then applying Theorem 6 to one of the negative edges gives

$$
\begin{aligned}
W(C_{r,s}) &= x_- W(C_{r,s-1}) + y_- W(P_{r,s-1}) \\
&= x_- W(C_{r,s-1}) + y_- X_+^r X_-^{s-1}.
\end{aligned}
$$

Hence, by induction,

$$W(C_{r,s}) = \sum_{i=0}^{s-1} x_-^i y_- X_-^{s-i-1} X_+^r + x_-^s W(C_{r,0}),$$

for all $s \geqslant 0$. If $r \geqslant 1$, combining this with (3.17) gives

$$W(C_{r,s}) = \sum_{i=0}^{s-1} x_-^i y_- X_-^{s-i-1} X_+^r + \sum_{i=0}^{r-2} x_-^s x_+^i y_+ X_+^{r-i-1} + x_-^s x_+^{r-1} Y_+. \qquad (3.18)$$

We could also obtain this formula directly from the definition of W, as follows. We take the order ϕ such that all positive edges precede all negative edges. The spanning trees of $C_{r,s}$ are obtained by omitting one edge e, and this edge will be externally active if and only if it is this first edge in the whole cycle. Thus we have a factor of Y_+ in the last term, corresponding to omitting the first edge, and we have $r-1$ other terms containing a factor of y_+, corresponding to omitting some other positive edge, and s terms containing a factor of y_-, corresponding to omitting a negative edge. An edge f in the tree is internally active if and only if it precedes e in the order. In (3.18) we have written each term as a product of the weights of the edges taken in the reverse order to ϕ. Thus we have factors of X_+ or X_- to the right of the factor y_+ or y_-, and factors of x_+ or x_- to the left.

Note that we could calculate $W(C_{r,s})$ differently, by applying Theorem 6 to the edges in some other order, or by taking a different order ϕ above. For example, applying Theorem 6 to the positive edges first, we would obtain

$$W(C_{r,s}) = \sum_{i=0}^{r-1} x_+^i y_+ X_+^{r-i-1} X_-^s + \sum_{i=0}^{s-2} x_+^r x_-^i y_- X_-^{s-i-1} + x_+^r x_-^{s-1} Y_-. \qquad (3.19)$$

At first sight, equations (3.18) and (3.19) appear to be inconsistent. However, we are always working in a quotient ring satisfying the conditions of Corollary 4. In this case, Corollary 4 guarantees that the two forms of $W(C_{r,s})$ calculated are equal.

3.7. Adding colours
When considering weighted graphs, it is natural to consider the operation of replacing two parallel edges with weights w_1 and w_2 by a single edge of weight $w_1 + w_2$. Here we have colours rather than weights, but we can use the recurrence relations established in Section 3.5 to show that, for a suitable choice of colour, replacing two edges by one does not affect the polynomial W.

Theorem 7. *Let (G, c) be a coloured graph with two parallel edges, e_1 and e_2, of colours λ and μ respectively. Let (G', c') be the graph formed from G by deleting these edges and replacing them by a single edge e of colour ν. Then if $I \subseteq \mathbb{Z}_{\Lambda,\alpha_i}$ is an ideal with $I_0' \subseteq I$ such that the equation of 4-tuples*

$$(X_\nu, Y_\nu, x_\nu, y_\nu) = (X_\lambda y_\mu + Y_\lambda x_\mu, Y_\lambda Y_\mu, x_\lambda y_\mu + Y_\lambda x_\mu, y_\lambda y_\mu) \qquad (3.20)$$

holds in $\mathbb{Z}_{\Lambda,\alpha_i}/I$, we have $W(G, c) = W(G', c')$ in $\mathbb{Z}_{\Lambda,\alpha_i}/I$.

Proof. When e is a loop, that is, e_1 and e_2 are both loops, we have from Theorem 6 that $W(G) = Y_\lambda Y_\mu W(G - e_1 - e_2) = Y_\nu W(G' - e) = W(G')$, where we have suppressed the dependence on c or c' for conciseness. When e is a bridge, we have that neither e_1 nor e_2 is a bridge, but that when one of them is deleted, the other becomes a bridge.

Applying Theorem 6 to G, first to e_2 and then to e_1, shows that $W(G) = x_\mu Y_\lambda W(G/e_2 - e_1) + y_\mu X_\lambda W(G - e_2/e_1)$. However, since $G/e_2 - e_1 = G - e_2/e_1 = G'/e$, this reduces to $W(G) = (x_\mu Y_\lambda + y_\mu X_\lambda) W(G'/e)$. However, from (3.20), this is just $X_v W(G'/e) = W(G')$. If e is neither a bridge nor a loop the argument is similar – applying Theorem 6 to e_1 and e_2, we express $W(G)$ in terms of $W(G'/e)$ and $W(G' - e)$, and the coefficients x_v, y_v are such that applying Theorem 6 again shows that $W(G) = W(G')$. □

We call a colour v whose variables satisfy (3.20) the *parallel sum* of the colours λ and μ. In the special case when the variables are of the form $X_c = t + x_c$, $Y_c = 1 + zx_c$, $y_c = 1$, the sum formula was given by Traldi [32].

Having shown how colours can be 'added', it is natural to ask whether there is a 'zero' colour, such that the presence or absence of edges of this colour does not affect $W(G)$. As it stands the answer to this question is 'no'. For example, consider the graph (G, c) with n vertices and one edge, of colour 0. This graph has $W(G, c) = X_0 \alpha_{n-1}$. Since the α_i are independent, there is no choice of X_0 for which this will always be equal to $W(E_n) = \alpha_n$. This is an example of a situation where we can obtain some additional property of W by insisting that the α_n are of the form $C\alpha^{n-1}$.

Theorem 8. *Let $f : \mathbb{Z}_{\Lambda, \alpha_i}/I_0' \to R$ be a ring homomorphism, with $f(\alpha_n) = C\alpha^{n-1}$ for some $C, \alpha \in R$. Suppose in addition that $f(X_0) = \alpha$, $f(Y_0) = 1$, $f(x_0) = 0$ and $f(y_0) = 1$. Then, for any coloured graph (G, c) and edge $e \in E$ of colour 0, we have*

$$f(W(G, c)) = f(W(G - e, c)).$$

Proof. As before, we split into cases according to whether e is a bridge, a loop, or neither. If e is a bridge, then we have that $f(W(G, c)) = \alpha f(W(G/e, c))$. Now G/e and $G - e$ have the same blocks, but $G - e$ has one more component. The result thus follows from (3.15). In the remaining cases, the result follows immediately from Theorem 6. □

Note that the zero colour is indeed a zero for the addition of colours described in Theorem 7. Note also that under the conditions of Theorem 8, we can add edges of colour 0 between all vertices of a graph G not already connected, without changing $W(G)$. We can then use Theorem 7 to replace any parallel edges by single edges of a new colour. This gives a complete graph which may have some loops attached to some of its vertices. Since the effect of a loop of colour λ is to multiply W by Y_λ, we can replace these by a single loop of an appropriate colour. Thus, if we like, any graph can be considered as a complete graph with one loop, for the purpose of calculating $W(G)$.

In this section we have so far considered merging two edges in parallel. In the planar case this operation has a dual – merging two edges e_1 and e_2 having a common end vertex v of degree exactly 2. We say such edges are *in series*. This operation can be applied to any graph, and we have results corresponding to Theorem 7 and Theorem 8. In the planar case we can appeal to (3.2) for their proofs, but in any case the proofs are similar to those of Theorems 7 and 8, so we shall omit them.

Theorem 9. *Let (G, c) be a coloured graph with two edges e_1 and e_2 in series, where $c(e_1) = \lambda$ and $c(e_2) = \mu$, and e_1, e_2 are not loops. Let (G', c') be the graph formed from G by replacing these edges with a single edge of colour v. Then, if $I \subseteq \mathbb{Z}_{\Lambda, \alpha_i}$ is an ideal with $I'_o \subseteq I$ such that the equation of 4-tuples*

$$(X_v, Y_v, x_v, y_v) = (X_\lambda X_\mu, Y_\lambda x_\mu + X_\lambda y_\mu, x_\lambda x_\mu, y_\lambda x_\mu + X_\lambda y_\mu)$$

holds in $\mathbb{Z}_{\Lambda, \alpha_i}/I$, we have $W(G, c) = W(G', c')$ in $\mathbb{Z}_{\Lambda, \alpha_i}/I$.

Also, if (G, c) is any coloured graph, $I \subseteq \mathbb{Z}_{\Lambda, \alpha_i}$ is any ideal containing I_0 and e is an edge of G with colour 1, where

$$(X_1, Y_1, x_1, y_1) = (1, 1, 1, 0),$$

then $W(G, c) = W(G/e, c)$ in $\mathbb{Z}_{\Lambda, \alpha_i}/I$.

We call a colour v as above the *series sum* of the colours λ and μ, and note that a colour 1 as above is a zero for series addition.

3.8. A comment on the noncommutative case

It is natural to ask why, throughout this section, we have considered only maps into *commutative* rings, rather than into quotients of the free polynomial algebra generated over $\mathbb{Z}$ by $\{X_\lambda, Y_\lambda, x_\lambda, y_\lambda : \lambda \in \Lambda\}$. The answer is that, in the noncommutative case, there is a difficulty in choosing an order in which to multiply the edge weights to obtain the weight of a spanning tree. One possibility is to use the same order ϕ used for deciding the edge activities. The problem with this is that, in our proof of the order independence of W, we assume that the weight of a tree does not change when we change the order of two consecutive, nonrelated edges e and f. Since no edge activities or weights change, this is true provided that the weights of e and f commute. These edges could have any possible weights, so we can obtain nothing new in the noncommutative case without substantially modifying this proof.

Another possibility is to use a fixed order ψ, independent of ϕ, with the aim of defining a function $f(G, c, \psi, \phi)$, depending on ψ but not ϕ. In this case, the proof in Section 3.2 will go through with minor modifications, though the conditions required will be more complicated. For example, condition (3.3) will be replaced by

$$X_\lambda w y_\mu - y_\lambda w X_\mu - x_\lambda w Y_\mu + Y_\lambda w x_\mu \in I,$$

where w is any product of edge weights, in any order. Unfortunately, in this case we have a problem with the recurrence relations. For an edge e not the first or last in the order ψ, calculating $f(G)$ from $f(G/e)$ and $f(G - e)$ involves the rather peculiar operation of inserting a factor of $x_{c(e)}$ or $y_{c(e)}$ into a particular position in all the products involved. Thus, while we could use a spanning tree expansion to define a function $f(G, c, \psi)$ taking values in a noncommutative ring, it would not obey simple recurrence relations, and we shall restrict our attention to the commutative case. Nevertheless, we would not like to rule out the possibility of making use of a noncommutative generalization of the Tutte polynomial at some point in the future.

This concludes our description of a Tutte polynomial for coloured graphs. In the next two sections we shall describe two weighted-graph polynomials that have been defined in different ways, and show that they can be obtained from the polynomial W defined here.

4. Random subgraph definition

4.1. The random cluster model of Fortuin and Kasteleyn

In [7], Fortuin and Kasteleyn define a polynomial on weighted graphs as follows. Given a graph G, a real number p_e for each edge e of G, and one extra real number κ,

$$Z(G, (p_e), \kappa) = \sum_{F \subseteq E(G)} \left(\prod_{e \in F} p_e \right) \left(\prod_{e \in E - F} q_e \right) \kappa^{k(F)},$$

where $q_e = 1 - p_e$, and $k(F)$ is the number of components of F, considered as a spanning subgraph of G. In the case where each p_e lies between 0 and 1, we can consider F as a random subgraph of G, where the edges are selected independently, with e selected with probability p_e. In this case $Z(G, (p_e), \kappa)$ is just the expectation of $\kappa^{k(F)}$. Note that in the above definition we may take κ and p_e to be elements of any ring, and not necessarily real numbers. We shall show that the resulting polynomial Z can be obtained from the coloured Tutte polynomial W (defined in the previous section) by suitable substitutions.

In [7], Fortuin and Kasteleyn show that Z satisfies

$$Z(E_n) = \kappa^n, \tag{4.1}$$

and the relation

$$Z(G) = p_e Z(G/e) + q_e Z(G - e), \tag{4.2}$$

for any edge e of G. Clearly there can be only one polynomial satisfying (4.1) and (4.2), so if we can find some choice of variables for which W satisfies these equations, this evaluation of W will be equal to Z. Now taking $\alpha_n = \kappa^n$ gives $W(E_n) = \kappa^n$, as required. Considering the weights p_e as colours and taking $x_\lambda = \lambda$, $y_\lambda = 1 - \lambda$, we have from Theorem 6 that W satisfies (4.2) when e is neither a bridge nor a loop. Note that for a loop $G/e = G - e$ and for a bridge $W(G - e) = \alpha_{k(G)+1} W(G/e)/\alpha_{k(G)}$. Hence, taking $Y_\lambda = 1$ and $X_\lambda = \lambda + \kappa(1 - \lambda)$, we have from Theorem 6 that W satisfies (4.2) in all cases. It is straightforward to check that these choices satisfy (3.6), so that there is a well-defined evaluation of W satisfying (4.1) and (4.2). As noted above, this instance of W is thus precisely Z.

In fact it is easy to see that, in certain cases, W can be obtained from Z. On the one hand, if we restrict our attention to graphs where all the edges have the same colour or weight, that is, to unweighted graphs, then W is just the usual Tutte polynomial, while Z is equivalent to the rank generating function. As mentioned in the Introduction, these can be obtained from each other in a simple way. On the other hand, if we consider graphs where all the edges have different colours, then, as we can set each p_e to 0 or 1 independently, the polynomial Z contains all the information about how many components G has after certain edges are contracted and the rest deleted. It is straightforward to check that this information allows the calculation of $W(G)$ from its recurrence relations.

4.2. A slight extension of Z

We return to the interpretation of $Z(G,(p_e),\kappa)$ as the expectation of $\kappa^{k(F)}$, where F is a random subgraph of G, with edges selected independently with probabilities p_e. Setting $\kappa = 0$ in the polynomial $\kappa^{-1}Z$, which has no negative powers of κ, gives the probability that F is connected. Can we get the probability that F has exactly r components? The answer is 'yes': we can define $Z'(G,(p_e),(\kappa_s)_{s=1}^{\infty})$ by replacing $\kappa^{k(F)}$ by $\kappa_{k(F)}$ in the definition of Z. Now, setting $\kappa_r = 1$ and $\kappa_s = 0$ for $s \neq r$ gives a polynomial in (p_e) which gives the probability that F has exactly r components. There is a cost, however. While Z' still obeys (4.2), it no longer has a spanning tree expansion. This is because the variable X_λ in the expansion of Z is equal to $\lambda + \kappa(1 - \lambda)$, and multiplying by κ makes no sense for Z'. On the other hand, all the information contained in Z' is given by Z. In particular, we can calculate Z' by evaluating Z with κ as a free variable, using its spanning tree expansion if we wish, and then replacing κ^r by κ_r.

5. Rank generating function definition

As mentioned in the Introduction, the Tutte polynomial $T(G;x,y)$ is simply related to the dichromatic polynomial $Q(G;t,z)$, which is essentially the rank generating function of G. In [32], Traldi defines a dichromatic polynomial for graphs G with weights w_e on the edges as follows:

$$Q(G;t,z)(w_e)_{e\in E(G)} = \sum_{F\subseteq E} \left(\prod_{e\in F} w_e\right) t^{k(F)} z^{n(F)}, \tag{5.1}$$

where $k(F)$ and $n(F)$ are the number of components and the nullity of F, as before. This can also be considered as generalizing the rank generating function $R(G;w,z)$ for unweighted graphs, defined by (2.3). As in the unweighted case, whether we take $t^{k(F)}$ or $w^{r(F)}$ in (5.1) makes no difference: since $r(F) = |V(G)| - k(F)$, one polynomial can be obtained from the other by substituting $t = w^{-1}$ and multiplying by $w^{|V(G)|}$.

Now, in the unweighted case there is a difference between the rank generating function, which is a polynomial in two variables, and the random cluster model, which is a polynomial in only one variable. However, in the weighted case this difference disappears. In particular, since $n(F) = |F| - |V(G)| + k(F)$,

$$Q(G;t,z)(w_e)_{e\in E} = z^{-|V(G)|} \sum_{F\subseteq E} \left(\prod_{e\in F} zw_e\right)(tz)^{k(F)}$$

$$= z^{-|V(G)|} Q(G;tz,1)(zw_e)_{e\in E}.$$

This shows that the polynomial Q can be recovered from its evaluation with $z = 1$.

Finally, the substitutions $w_e = \frac{p_e}{1-p_e}$ and $p_e = \frac{w_e}{1+w_e}$ convert between $Q(G;t,1)$ and Z as follows:

$$Q(G; t, 1)(w_e)_{e \in E} = \sum_{F \subseteq E} \left(\prod_{e \in F} w_e \right) t^{k(F)}$$

$$= \left(\prod_{e \in E} 1 + w_e \right) \sum_{F \subseteq E} \left(\prod_{e \in F} \frac{w_e}{1 + w_e} \right) \left(\prod_{e \in E - F} \frac{1}{1 + w_e} \right) t^{k(E)}$$

$$= \left(\prod_{e \in E} 1 + w_e \right) Z(G, (p_e), t).$$

In summary, we have shown that the random cluster model of Fortuin and Kasteleyn [7] is equivalent to the generalization of the rank generating function introduced by Traldi [32], and that both can be obtained from the polynomial W, defined by generalizing the spanning tree expansion of the Tutte polynomial as far as possible.

6. Applications to links

In the mid-1980s, Jones [11] discovered a new and very powerful polynomial invariant of knots and links. Related polynomial invariants were later discovered by Kauffman [15] and by four groups working independently [8]. The second of these, the Homfly polynomial, is the universal link invariant satisfying certain linear relations, called *skein* relations, with arbitrary coefficients (see [8] or [6]). For relations of different forms, much effort has been spent on finding the most general polynomial invariants satisfying such conditions (for example, [13], [17], [18], [23]). Soon it was shown that the Jones polynomial and the Kauffman bracket are closely related to the Tutte polynomial [29]; this relationship is frequently expressed in terms of a 'state model', or a rank generating function. A more general link invariant, the two-variable Kauffman polynomial, is also related to the Tutte polynomial. These connections were used by Kauffman [13], Murasugi [19, 20], and Thistlethwaite [29, 30] to answer several questions about links, including long-standing conjectures of Tait [28]. For a review of many of these results, see [16].

Our aim in this section is to find the most general link invariant under different conditions from all those previously considered. We demand that our link invariant have a spanning tree expansion, that is, we look for the most general link invariant that can be obtained from the coloured Tutte polynomial W. For this reason, we wish to use graphs to represent links. A different connection between graphs and links has been used by Jaeger [9] and by Traldi [32], to relate the Tutte polynomial and the Homfly polynomial. However, this works the other way round, relating the Tutte polynomial of any planar graph to the Homfly polynomials of certain special links.

6.1. The relationship between links and signed graphs

In this section we shall describe a (non-unique) way of associating a signed plane graph to a link diagram, such that the link diagram can be recovered from the graph. We shall then give precise 'if and only if' conditions for two graphs to represent the same link.

A *link diagram* D is a union of finitely many smooth closed curves in $S^2 = \mathbb{R}^2 \cup \{\infty\}$ which has finitely many transverse double points, called *crossings*, and no other multiple

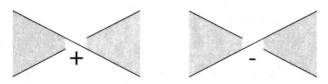

Figure 1   The sign of an edge associated with a crossing.

points, together with information specifying which arc is on top at each crossing. In other words, a link diagram is a 4-regular graph drawn in S^2, possibly with some extra simple closed curves, together with crossing information.

By a *shaded link diagram* we mean a link diagram D with alternate regions shaded; there are two shaded link diagrams for each D.

Given a connected link diagram D, which we may as well assume to lie in $\mathbb{R}^2$, we can associate a connected 2-coloured plane graph to each shading of D as follows. Take a vertex for each shaded region and an edge for each crossing, joining the two shaded regions which meet there. Colour each edge $+$ or $-$ according to the sense of the crossing, as shown in Figure 1. For disconnected diagrams, there is a complication. For example, shading a diagram with n components and no crossings will give anywhere from 1 to n shaded regions, depending on the arrangement of the components. However, we would like each component with no crossings to be represented by a vertex, to avoid losing information. Thus, for a disconnected diagram, we first shade the diagram, and then take the disjoint union of the graphs corresponding to each component, with the induced shading. We shall say that two shaded diagrams are *equivalent* if we can pair off their components such that corresponding components are related by an isotopy of S^2. Note that we require corresponding components to have the same shading. Note also that there is only one shading of a loop with no crossings, as we are working in S^2. Similarly, two signed graphs drawn in $\mathbb{R}^2 = S^2 - \{\infty\}$ are equivalent if their components are related by isotopies of S^2. We shall write $\mathscr{P}_s$ for the set of equivalence classes of signed plane graphs. Note that we have a one-to-one correspondence between equivalence classes of shaded diagrams and elements of $\mathscr{P}_s$. From now on we shall regard equivalent graphs or diagrams as being the same.

Now that we have a way of associating graphs to link diagrams, we would like to know when two graphs represent the same link, so that we can give conditions under which a signed plane graph invariant gives rise to a link invariant.

From [24], or [5], we know that two link diagrams represent the same link if and only if they are related by a series of Reidemeister moves. Now Reidemeister moves have a natural interpretation on shaded diagrams, and hence on signed plane graphs. We shall refer to these operations on graphs as *graph Reidemeister moves*, or *GR-moves*. Since these operations are of central importance to defining link polynomials via graphs, we first list them precisely, and then show the correspondence with Reidemeister moves. In describing these moves, we shall call two parallel edges of a plane graph *adjacent* if they bound a face.

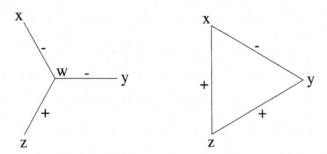

Figure 2   The star-delta transformation.

Two signed plane graphs, that is, elements of $\mathscr{P}_s$, are *related by a GR-move* if one can be obtained from the other by one of the following operations:

1a. replacing G by G/e, where e is a pendant edge, that is, an edge having an end-vertex of degree 1, and so being a bridge

1b. replacing G by $G - e$, where e is a loop

2a. replacing G by $G - e_1 - e_2$, where e_1 and e_2 are adjacent parallel edges, with opposite sign, including the case where e_1 and e_2 are loops

2b. replacing G by $G/e_1/e_2$, where e_1 and e_2 are edges of opposite sign, which are in series (*i.e.*, share a vertex of degree 2), but are not parallel

2c. replacing G by the disjoint union of G_1 and G_2, where G has edges e_1 and e_2 of opposite sign between distinct vertices x and y, where x has degree 2, and the closed curve $e_1 \cup e_2$ splits $G - x$ into subgraphs G_1 and G_2 meeting at y

3a. the star-delta transformation of replacing the first configuration shown in Figure 2 by the second, or

3b. as 3a, but with the signs of all edges changed.

Note that in the star-delta transformation of 3a and 3b, the vertices x, y and z need not be distinct. However, in the second configuration the three (distinct) edges shown must form the boundary of a face.

In each of the moves 1a and 1b we shall distinguish two cases, according to the sign of e, so we shall consider GR-moves $1a^+$, $1a^-$, $1b^+$ and $1b^-$.

We can now state precisely the relationship between Reidemeister moves and GR-moves.

Lemma 10.   *Two shaded link diagrams L_1 and L_2 are related by a shading-respecting Reidemeister move if and only if the corresponding graphs G_1 and G_2 are related by a GR-move.*

Proof.   We consider each Reidemeister move in turn, showing that two shaded link diagrams are related by such a move (in either direction) if and only if the corresponding graphs are related by one of the corresponding GR-moves (in either direction).

1. The first Reidemeister move: removing a kink. There are four cases, depending on the sign of the kink, and the shading of the graph. We start with a positive kink, shaded as

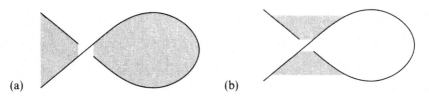

Figure 3 The two shadings of a positive kink.

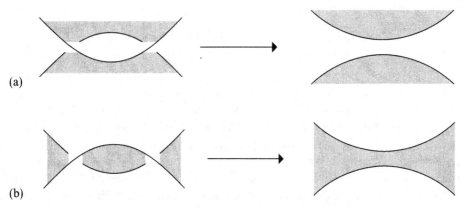

Figure 4 The two shadings of the second Reidemeister move.

in Figure 3(a). Now the shaded diagrams L containing such a kink are precisely those corresponding to graphs G with a negative pendant edge e. Also, removing the kink gives the diagram L' corresponding to the graph $G' = G/e$. Thus two shaded diagrams are related by this case of Reidemeister 1 if and only if the corresponding graphs are related by GR-move $1a^-$, *i.e.*, contracting a negative pendant edge.

Next we consider the other shading of a positive kink, which is shown in Figure 3(b). This time, removing the kink corresponds to GR-move $1b^+$, that is, deleting a positive loop. Similarly, removing a negative kink corresponds to GR-move $1a^+$, that is, contracting a negative pendant edge, or to GR-move $1b^-$, deleting a negative loop. Thus two shaded diagrams are related by Reidemeister move 1 if and only if the corresponding graphs are related by GR-move $1a$ or $1b$.

2. The second Reidemeister move. In this case we must be careful, as the number of components of the link diagram may change. We start by considering the shading shown in Figure 4(a). This corresponds to removing a pair of adjacent parallel edges e_1 and e_2 (which may be loops) with opposite sign, which is exactly GR-move $2a$. Note that a component of the link diagram becomes disconnected under this move if and only if the two shaded regions involved are distinct, and are not connected at other crossings via a sequence of other shaded regions. This is the same condition as e_1 and e_2 having distinct end points, and their removal disconnecting these vertices. Thus the link diagram and the

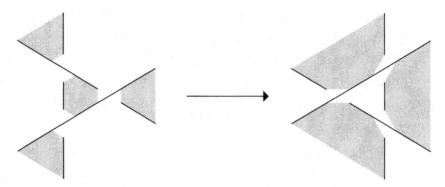

Figure 5   A shaded version of the third Reidemeister move.

graph gain a component in the same circumstances, and the association between possibly disconnected diagrams and possibly disconnected graphs described above is preserved.

We now consider the other shading of Reidemeister 2, shown in Figure 4(b), corresponding to contracting a positive edge e_1 and a negative edge e_2 in series, that is, sharing a vertex of degree 2. We have a correspondence between this Reidemeister move, *restricted to the case where e_1 and e_2 are not parallel,* and GR-move 2b. Note that under this restriction, which corresponds to the two outer shaded regions in the picture being distinct, the number of components of both the diagram and the graph remains the same. If e_1 and e_2 are also parallel, then the diagram becomes disconnected. The vertex representing the outer region must thus be replaced by two vertices, one for each of the two new components. This is precisely GR-move 2c. Thus, two shaded link diagrams are related by this shading of Reidemeister move 2, if and only if the corresponding graphs are related by GR-move 2b or 2c.

3. The third Reidemeister move. This time it turns out that there are fewer cases than we expect. We start by considering the first version of Reidemeister 3, moving a string underneath a crossing, from one B-region to the other. One shading is shown in Figure 5: this corresponds to GR-move 3a. Next we consider the other shading of the same Reidemeister move. Since this is just Figure 5 backwards, it also corresponds to GR-move 3a. Finally, the two shadings of the other version of Reidemeister move 3, that is, Figure 5 with all crossings reversed, correspond to GR-move 3b.

This completes the proof of the lemma.   □

From [24], or [5], two graphs G_1 and G_2 represent the same link if and only if the corresponding *unshaded* link diagrams L_1 and L_2 are related by a sequence of Reidemeister moves. However, to make use of the lemma above, we must consider shaded diagrams. Now, if L_1 and L_2 are related as unshaded diagrams, and we perform the corresponding sequence of GR-moves on G_1, we may not obtain the graph G_2. This is because there is another graph G_2' corresponding to the other shading of L_2. However, as we shall show below, the two graphs obtained from the two shadings of any diagram are always related

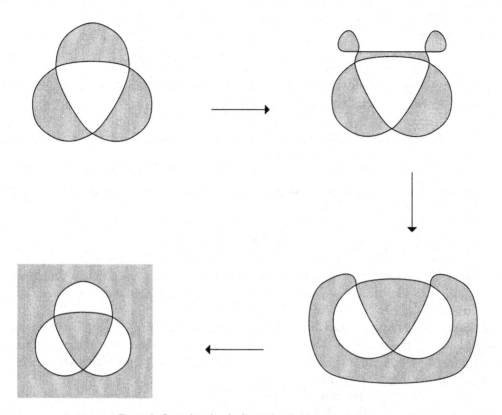

Figure 6   Inverting the shading using Reidemeister moves.

by a sequence of GR-moves. Combined with Lemma 10 above and the result from [5], this will show that two graphs represent the same link if and only if they are related by a sequence of GR-moves.

Lemma 11.   *Let L be any link diagram, and let L_1 and L_2 be the two shaded link diagrams corresponding to L. Then the corresponding graphs, G_1 and G_2, are related by a sequence of GR-moves.*

Proof.   Given Lemma 10, it suffices to show that L_1 and L_2 are related by a sequence of *shading-respecting* Reidemeister moves. This is true because we can take any string in L_1, and use Reidemeister moves to pull part of this string across the whole diagram, past the point at infinity, and back to where it started. This procedure is illustrated for a trefoil by Figure 6.

Note that at the end of this procedure we have the same link diagram, as we have put the string back where it started. On the other hand, as each crossing has had a string pulled over it using Reidemeister move 3, the shading near each crossing has changed. As there are only two ways of shading a diagram, the shading of the whole diagram must have been inverted.   □

As noted above, combining Lemmas 10 and 11 with the result from [24], or [5], that link diagrams represent the same link if and only if they are related by a sequence of Reidemeister moves, we have the following corollary.

Corollary 12. *Two signed plane graphs* $G_1, G_2 \in \mathscr{P}_s$ *represent the same link if and only if they are related by a sequence of GR-moves.*   □

With the aid of this corollary, we can now establish under which conditions the coloured Tutte polynomial W is an invariant of links.

6.2. A link polynomial from the coloured Tutte polynomial

The association between link diagrams and graphs described in the previous section provides a non-unique way of evaluating the polynomial W on a link diagram, and hence on a link. Since we shall have to relate graphs with different numbers of components, we shall use the form of W satisfying $W(E_n) = C\alpha^{n-1}$.

In fact, this choice is forced on us as described in Section 3.7 – writing 0 for the parallel sum of the colours $+$ and $-$, GR-move 2a invariance says that the polynomial is unchanged by deleting edges of colour 0, so considering the graph with n vertices and one edge shows that we must have $\alpha_n = X_0\alpha_{n-1}$.

We wish to establish the conditions on $C, \alpha, X_{\pm}, Y_{\pm}, x_{\pm}, y_{\pm}$ necessary to make this evaluation on links well defined. Recalling that the Kauffman bracket itself is only invariant under Reidemeister moves 2 and 3, but can be normalized to be fully Reidemeister invariant, we shall consider instead a suitably normalized version of W. To do this we need the concept of the *writhe* of an oriented link diagram.

It is easily seen that in an oriented diagram the crossings can be classified into two types, depending on the relationship between the orientation of the 'overpass' to that of the 'underpass'. The writhe simply counts the number of one type minus the number of the other (see [15] for the appropriate diagrams). The relevant properties of the writhe are that it is unchanged by Reidemeister moves 2 and 3, and that it decreases by one when a positive kink is removed, and increases by one when a negative kink is removed. It is sometimes stated that, since the writhe is only defined for oriented links, polynomials such as the Kauffman bracket can only be normalized to give invariants of oriented links. In fact, as Kauffman himself was aware [15], the *self-writhe* works just as well, and makes sense for unoriented links. The self-writhe of an oriented link diagram is defined in the same way as the writhe, but counting only the crossings involving only one component of the link. The self-writhe changes in exactly the same way as the writhe under Reidemeister moves, but does not actually depend on the orientation, and can thus be defined for unoriented diagrams. For our purposes, there is no reason to introduce orientation, so we shall work with the self-writhe.

Setting $a(G)$ to be the number of positive edges of G, $b(G)$ the number of negative edges, $c(G)$ the number of vertices, $d(G)$ the number of components, and $w(G)$ the self-writhe of the link diagram associated to G, our aim is to find the most general map S of the form

$$S : \mathscr{P}_s \longrightarrow \mathbb{Z}[a, b, c, d, w, w^{-1}, C, \alpha, X_+, Y_+, x_+, y_+, X_-, Y_-, x_-, y_-]/I,$$

for some ideal I, where

$$S(G) = a^{a(G)} b^{b(G)} c^{c(G)} d^{d(G)} w^{w(G)} W(G),$$

and $S(G)$ depends only on the link represented by G. In fact, we would like a map S such that every other such map S' is of the form $S' = qS$, where q is a quotient map $A/I \longrightarrow A/J = (A/I)/(J/I)$. However, we do not wish to keep 'dummy' variables, namely, variables whose presence does not give us a more general polynomial.

Before continuing, we remark that if our only aim is to obtain a link invariant from $S(G)$, then we do not have to quotient by an ideal I. In general, it suffices to find some equivalence relation $\sim$ such that $S(G_1) \sim S(G_2)$ whenever G_1 and G_2 represent the same link. However, for the resulting invariant to be of any use, the equivalence relation $\sim$ should be reasonably simple. Here we have chosen to consider only relations corresponding to quotienting by an ideal, but some other choice may be possible.

As we shall see, the ideal I we shall find is unchanged under the simultaneous substitutions $a \mapsto 1$, $X_+ \mapsto aX_+$, $Y_+ \mapsto aY_+$, $x_+ \mapsto ax_+$ and $y_+ \mapsto ay_+$. Also, the definition of W implies that $S(G)$ can be recovered from its evaluation at $a = 1$:

$$S(G)(a, b, c, d, w, w^{-1}, C, \alpha, X_+, Y_+, x_+, y_+, X_-, Y_-, x_-, y_-)$$
$$= S(G)(1, b, c, d, w, w^{-1}, C, \alpha, aX_+, aY_+, ax_+, ay_+, X_-, Y_-, x_-, y_-).$$

This holds since each positive edge contributes one factor of either X_+, Y_+, x_+ or y_+ to the weight of each spanning tree. Similar arguments show that $S(G)(a, b, c, d, \ldots)$ can be recovered from its evaluation at $a = b = c = d = 1$, using the substitutions $b \mapsto 1$, $X_- \mapsto bX_-$, $Y_- \mapsto bY_-$, $x_- \mapsto bx_-$, $y_- \mapsto by_-$, followed by $d \mapsto 1$, $\alpha \mapsto d\alpha$, $C \mapsto dC$, and $c \mapsto 1$, $X_+ \mapsto cX_+$, $X_- \mapsto cX_-$, $x_+ \mapsto cx_+$, $x_- \mapsto cx_-$, $C \mapsto cC$. Finally, to fix normalization, we take $S(E_1) = 1$, that is, $C = 1$. Again, this is just eliminating a dummy variable, as $S(G)$ has a factor of C for every graph $G \in \mathscr{P}_s$.

For the rest of this section, we thus confine our attention to this restriction of S, which we consider as a map

$$S : \mathscr{P}_s \longrightarrow \mathbb{Z}[w, w^{-1}, \alpha, X_+, Y_+, x_+, y_+, X_-, Y_-, x_-, y_-]/I,$$

given by $S(G) = w^{w(G)} W(G)$.

We now wish to establish necessary and sufficient conditions on the ideal I to ensure that $S(G)$ is well defined on links. More precisely, we shall prove the following result.

Theorem 13. *Given an ideal $I \subseteq \mathbb{Z}[w, w^{-1}, \alpha, X_\pm, Y_\pm, x_\pm, y_\pm]$, the map*

$$S : \mathscr{P}_s \longrightarrow \mathbb{Z}[w, w^{-1}, \alpha, X_\pm, Y_\pm, x_\pm, y_\pm]/I$$

defined by $S(G) = w^{w(G)} W(G)$ gives rise to a well-defined link invariant if and only if the following relations hold in $\mathbb{Z}[w, w^{-1}, \alpha, X_\pm, Y_\pm, x_\pm, y_\pm]/I$:

$$\begin{aligned} X_+ &= Y_- = w \\ X_- &= Y_+ = w^{-1}, \end{aligned} \tag{6.1}$$

$$\alpha = X_- y_+ + Y_- x_+, \tag{6.2}$$

$$X_\epsilon = x_\epsilon + \alpha y_\epsilon \tag{6.3}$$

for $\epsilon = +, -$, and

$$Y_\epsilon = \alpha x_\epsilon + y_\epsilon \tag{6.4}$$

for $\epsilon = +, -$.

Remark. Note that (6.3) and (6.4) imply that conditions (3.3)–(3.5) of Theorem 2 are satisfied, and hence (by Corollary 4) that $W(G)$ is well defined on graphs.

Before starting the proof, we note a consequence of equations (6.1)–(6.4) which will be used in the proof, and which is of interest in its own right. Combining (6.1) with (6.3) and (6.4), we have

$$\begin{aligned}
x_+ + \alpha y_+ &= y_- + \alpha x_- &= w, \\
\alpha x_+ + y_+ &= \alpha y_- + x_- &= w^{-1}.
\end{aligned} \tag{6.5}$$

Eliminating variables from these simultaneous equations, we obtain

$$(\alpha^2 - 1)(x_+ - y_-) = (\alpha^2 - 1)(y_+ - x_-) = 0. \tag{6.6}$$

This tells us that, if $\alpha^2 - 1$ is invertible, the 'dual variables' x_+ and y_- must be equal.

Proof of Theorem 13. We know from Corollary 12 that S is well defined on links if and only if it is invariant under GR-moves, so we consider each of these in turn.

1. GR-move 1a$^-$. Let G' be obtained from G by GR-move 1a$^-$, so $G' = G/e$, where e is a negative pendant edge. Then, from Theorem 6, we know that $W(G) = X_- W(G')$. Also, if L and L' are the corresponding link diagrams, then L' is obtained from L by removing a positive kink. Thus $w(L) = w(L') + 1$, and hence, by definition, $w(G) = w(G') + 1$. Thus $S(G) = wX_- S(G')$, and we have invariance under this GR-move provided $wX_- = 1$, that is, $X_- = w^{-1}$. This condition is also necessary, as can be seen by taking G to consist of two vertices joined by a negative edge.

2. GR-move 1b$^+$. This time let G' be obtained from G by deleting a positive loop, so $W(G) = Y_+ W(G')$. The operation on the link diagram is again removing a positive kink, so the writhe again decreases by one, and it is sufficient to have $wY_+ = 1$, or $Y_+ = w^{-1}$. As before, the case when G is just a positive loop shows this condition to be necessary as well.

3. GR-moves 1a$^+$ and 1b$^-$. These correspond to removing a negative kink, so the writhe increases by one. Arguing as for moves 1a$^-$ and 1b$^+$ shows that invariance under these moves is equivalent to requiring $X_+ = Y_- = w$.

To summarize so far, the polynomial S is invariant under the graph equivalents of Reidemeister 1 if and only if (6.1) holds. Unfortunately, for the remaining GR-moves the situation is more complicated, as the effect on the polynomial W depends on the rest of the graph.

4. GR-move 2a: removing a pair of adjacent parallel edges with opposite sign. The writhe does not change, so we require

$$W(G) = W(G - e_1 - e_2), \tag{6.7}$$

for all $G \in \mathscr{P}_s$ and pairs of edges $\{e_1, e_2\}$ as above. Now $W(G)$ does not depend on the embedding of G into the plane. Thus any conditions that make W invariant under GR-move 2a will also make W invariant under the more general operation of deleting any two parallel edges of opposite sign. However, since we are looking for necessary and sufficient conditions for GR-move 2a invariance, we must ensure that in the examples we use to show our conditions are necessary, the edges e_1 and e_2 really are adjacent.

Note that if e_1 and e_2 are loops, then (6.7) follows from (6.1). Thus we need only impose (6.7) for graphs in which e_1 and e_2 are not loops.

Our aim now is to replace the rather daunting set of conditions given by (6.7) by an equivalent, but much simpler, set of conditions. Let G, e_1, e_2 be as above, with G having another edge e_3, which is a positive loop. Set $G' = G - e_1 - e_2$. Then we have from Theorem 6 that

$$W(G) = Y_+ W(G - e_3), \text{ and } W(G') = Y_+ W(G' - e_3).$$

This implies that the relation (6.7) for G follows from (6.7) for the smaller graph $G - e_3$, so we do not need to impose (6.7) for G as a separate condition. A similar argument holds for graphs G containing negative loops, or bridges. Thus it suffices to demand (6.7) for graphs G containing neither loops nor bridges. From now on, we consider only such graphs G.

If G has no edges other than e_1 and e_2, then $W(G) = X_-y_+ + Y_-x_+$, while $W(G - e_1 - e_2) = W(E_2) = \alpha$, so (6.7) becomes

$$\alpha = X_-y_+ + Y_-x_+,$$

which is just (6.2).

If G has another edge e_3, the situation is more complicated. From Theorem 6, we have

$$W(G) = x_\epsilon W(G/e_3) + y_\epsilon W(G - e_3), \tag{6.8}$$

where ϵ is the sign of e_3. Also, *provided e_3 is not a bridge in $G' = G - e_1 - e_2$*, we have

$$W(G') = x_\epsilon W(G'/e_3) + y_\epsilon W(G' - e_3). \tag{6.9}$$

Now, given (6.8) and (6.9), the relation (6.7) for G again follows from (6.7) for smaller graphs, as contracting or deleting e_3 leaves e_1 and e_2 as parallel edges, though they may become loops. Hence, it suffices to consider (6.7) for graphs G such that every edge $e_3 \in E(G')$ is a bridge in G'. Thus G' is a forest which becomes 2-connected after the addition of the parallel edges e_1 and e_2. Hence G' has two end-vertices, which are precisely the vertices of e_1 and e_2, so that G consists of our two edges e_1 and e_2 between two distinct vertices x and y, and an x-y path, shown in Figure 7. Let us write $G_{r,s}$ for this graph G if the x-y path is a $P_{r,s}$ path, that is, one with r positive edges and s negative ones. Also, let us write $A_{r,s}$ for a path with one double edge consisting of edges of opposite sign, and $r + s$ single edges, of which r are positive (and s negative). In this notation, the additional

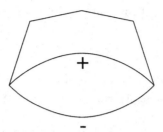

Figure 7 A family of graphs representing the unknot.

conditions we require for GR-move 2a invariance are just

$$W(G_{r,s}) = W(P_{r,s}), \tag{6.10}$$

for all nonnegative r, s with $r + s \geqslant 1$. From the conditions (6.1) and (6.2) we have already imposed, we have that $Y_+ Y_- = 1$, and that $W(A_{0,0}) = \alpha$. Thus, when $r + s = 1$, applying Theorem 6 to the edge of $G_{r,s}$ that is not e_1 or e_2, shows that (6.10) and (6.3) are equivalent. Equation (6.3) is thus necessary for the GR-move 2a invariance of S. In fact, with (6.2) it is also sufficient, as can be shown by induction on $r + s$. To see this, note first that

$$W(A_{r,s}) = X_+^r X_-^s \alpha = \alpha W(P_{r,s}). \tag{6.11}$$

Now suppose that $r + s > 1$, and also that $s > 0$. Applying Theorem 6 to a negative edge of $G_{r,s}$ other than e_1 or e_2, we obtain

$$
\begin{aligned}
W(G_{r,s}) &= x_- W(G_{r,s-1}) + y_- W(A_{r,s-1}) \\
&= x_- W(P_{r,s-1}) + y_- W(A_{r,s-1}) \\
&= (x_- + \alpha y_-) W(P_{r,s-1}) \\
&= X_- W(P_{r,s-1}) = W(P_{r,s}),
\end{aligned}
$$

with the second equality following from the induction hypothesis, the third from (6.11), and the fourth from (6.3). If $s = 0$, then we can apply Theorem 6 to a positive edge, and argue exactly as above. This completes the induction, showing that (6.10), and hence the invariance of S under GR-move 2a, follows from (6.2) and (6.3).

We have now shown that, given (6.1), the invariance of S under GR-move 2a is equivalent to (6.2) and (6.3).

5. GR-move 2b: contracting a positive edge e_1 and a negative edge e_2 in series. As before, we can assume there are no bridges or loops in the rest of the graph. This time, when we contract e_1 and e_2, another edge e_3 that was not a bridge will still not be a bridge. On the other hand, an edge that was not a loop may become a loop. Thus we need to consider exactly those graphs $G \in \mathscr{P}_s$ with no bridges and loops in which all edges become loops when e_1 and e_2 are contracted, i.e., graphs of the form shown in Figure 8.

If there are no edges other than e_1 and e_2, then invariance under contracting e_1 and e_2 follows from (6.1). Otherwise, the situation is exactly as for GR-move 2a – we need to impose the condition $W(G) = W(G/e_1/e_2)$ only for graphs G with one extra edge, that is,

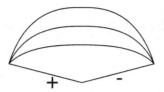

Figure 8 A second family of graphs representing the unknot.

to impose (6.4). Since the argument is exactly analogous to that for GR-move 2a above, we omit it here.

6. GR-move 2c. Let G and G' be related by GR-move 2c, and let e_1, e_2, x, y be as described in the definition of GR-move 2c. Note that y is a cut vertex in $G - e_1 - e_2$, so that G' and $G - e_1 - e_2$ have exactly the same blocks. Since they also have the same number of components, equation (3.15) tells us that $W(G') = W(G - e_1 - e_2)$. However, the argument for GR-move 2a shows that $W(G - e_1 - e_2) = W(G)$, given (6.1)–(6.3). As GR-move 2c does not change the writhe, the invariance of S under this move follows.

7. GR-move 3a: the star-delta transformation of replacing the first configuration in (6.1) by the second. In this case it will turn out that we need impose no extra conditions to ensure the invariance of W, and hence of S. Thus it makes no difference if we drop the constraint that the three edges shown in the second configuration of (6.1) bound a face.

Let $G, G' \in \mathcal{P}_s$ be signed plane graphs, such that G' can be obtained from G by GR-move 3a. As before, the writhe does not change, so for S to be invariant under this operation, we require

$$W(G) = W(G').\qquad(6.12)$$

If G has another edge e, then e will be a bridge/loop in G if and only if it is a bridge/loop in G', so we can use the recurrence relations (3.14) given by Theorem 6 to simplify (6.12). When we delete e in G and G' we certainly obtain a smaller instance of (6.12). In fact, when we contract e in both graphs this is also the case, as we are allowing some of the vertices x, y and z to be identified. As long as G has edges not shown in (6.1), we may thus use Theorem 6 to express (6.12) for G in terms of the same relation for smaller graphs, so we may assume that G has only the three edges shown. From now on, we consider only such graphs G.

Now, if x, y and z are all distinct, or if z is identified with exactly one of x and y, then G and G' are equivalent under GR-moves 1a–2c, so, assuming invariance of S under these moves, we do not need to impose (6.12) for these graphs. The same applies if x, y and z are all identified. This leaves us with the case where x and y are identified, but are distinct from z. In this case, G is the graph shown in Figure 9(a), and G' the graph in Figure 9(b). Assuming (6.1), we thus require W of a double positive edge to be the same as W of a double negative edge, that is,

$$X_+ y_+ + Y_+ x_+ = X_- y_- + Y_- x_-.\qquad(6.13)$$

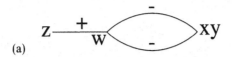

(a)

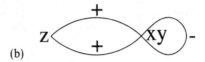

(b)

Figure 9 Two different graphs representing the Hopf link.

Given (6.3) and (6.4), this is equivalent to

$$\alpha x_+^2 + 2x_+ y_+ + \alpha y_+^2 = \alpha y_-^2 + 2y_- x_- + \alpha x_-^2. \tag{6.14}$$

However, from (6.5) and the fact that $ww^{-1} = 1$, we have

$$\alpha x_+^2 + (\alpha^2 + 1)x_+ y_+ + \alpha y_+^2 = 1 = \alpha y_-^2 + (\alpha^2 + 1)y_- x_- + \alpha x_-^2. \tag{6.15}$$

Now, subtracting (6.15) and (6.14), we see that the relation we require follows, as long as

$$(\alpha^2 - 1)x_+ y_+ = (\alpha^2 - 1)y_- x_-.$$

However,

$$(\alpha^2 - 1)x_+ y_+ - (\alpha^2 - 1)y_- x_- = (\alpha^2 - 1)(x_+ - y_-)y_+ + (\alpha^2 - 1)y_-(y_+ - x_-),$$

which is zero, from (6.6).

Combining the above observations, we see that (6.13), and hence invariance under GR-move 3a, follows from (6.1)–(6.4).

8. GR-move 3b. This operation is just GR-move 3a, but with all the signs changed. Following through the argument above leads to the same condition (6.13) for GR-move 3b invariance as for GR-move 3a invariance.

To conclude, the relations (6.1)–(6.4) are necessary and sufficient conditions for $S(G) = w^{w(G)}W(G)$ to be invariant under all GR-moves, and thus to give a well-defined link polynomial. This completes the proof of Theorem 13.   □

We remark that for the three-variable Kauffman bracket, as for S, the conditions for invariance under Reidemeister moves 1 and 2 imply invariance under Reidemeister move 3. In the Kauffman bracket case, there is a short proof, involving resolving a crossing, and applying Reidemeister 2 twice to one of the resulting diagrams. Unfortunately, that proof does not work for S, since the coefficients corresponding to resolving the crossing change from x_+, y_+ say, at the start of the Reidemeister move, to y_-, x_- at the end. As we do not wish to make any unnecessary assumptions, such as $x_+ = y_-$, a different proof of Reidemeister 3 invariance, such as the one given above, is necessary.

Writing J_0 for the ideal of $\mathbb{Z}[w, w^{-1}, \alpha, X_\pm, Y_\pm, x_\pm, y_\pm]$ generated by relations (6.1)–(6.4), Theorem 13 claims that the most general form of a link polynomial arising from the coloured Tutte polynomial is the map $S_0 : \mathscr{P}_s \longrightarrow \mathbb{Z}[w, w^{-1}, \alpha, X_\pm, Y_\pm, x_\pm, y_\pm]/J_0$, given by $S_0(G) = w^{w(G)}W(G)$. In the next section we shall show that S_0 can in fact be obtained from the three-variable Kauffman bracket, considered as a map into a suitable quotient ring.

6.3. A reduction of the link invariant S_0

So far we have attempted to describe the most general link invariant with a spanning tree expansion. As a consequence, the invariant S_0 we have produced has a rather complicated description: it is a map into $\mathbb{Z}[w, w^{-1}, \alpha, X_\pm, Y_\pm, x_\pm, y_\pm]$ quotiented by the relations listed in Theorem 13.

Now, considered as operations on link diagrams, contracting a positive edge and deleting a negative one both have the same effect, that is, resolving a crossing in such a way that the A-regions become joined. It is thus natural to set $x_+ = y_-$: as we shall see later, this loses no information about the link. In other words, we shall consider the composition of S_0 with a ring homomorphism ϕ such that

$$\begin{aligned} \phi(x_+) &= \phi(y_-) = A, \\ \phi(x_-) &= \phi(y_+) = B, \end{aligned} \tag{6.16}$$

and

$$\phi(\alpha) = d, \tag{6.17}$$

where the last choice is to make our notation consistent with [13]. Since we want Reidemeister invariance, from equations (6.1) and (6.3), we must have

$$\begin{aligned} \phi(w) &= \phi(Y_-) = \phi(X_+) = A + dB, \\ \phi(w^{-1}) &= \phi(Y_+) = \phi(X_-) = dA + B, \end{aligned} \tag{6.18}$$

We thus have all our variables given in terms of A, B, and d. In fact, with these substitutions, the polynomial W becomes the three-variable Kauffman bracket, as can be seen by considering the recurrence relations obeyed by each.

We would like to know under what conditions ϕ makes sense as a map from the quotient ring in which S_0 takes its values. This question is answered by the following result.

Theorem 14. *Let K be an ideal of $\mathbb{Z}[A, B, d]$, and let*

$$\phi : \mathbb{Z}[w, w^{-1}, \alpha, X_\pm, Y_\pm, x_\pm, y_\pm] \longrightarrow \mathbb{Z}[A, B, d]$$

be the ring homomorphism given by (6.16)–(6.18) above. Then ϕ induces a homomorphism

$$\overline{\phi} : \mathbb{Z}[w, w^{-1}, \alpha, X_\pm, Y_\pm, x_\pm, y_\pm]/J_0 \longrightarrow \mathbb{Z}[A, B, d]/K$$

if and only if $K \supseteq K_0$, where K_0 is the ideal generated by

$$(A + dB)(dA + B) = 1, \tag{6.19}$$

and

$$d = A^2 + 2dAB + B^2. \tag{6.20}$$

Proof. The map $\overline{\phi}$ will exist if and only if the relations given by the images of (6.1)–(6.4) hold in $\mathbb{Z}[A,B,d]/K$. Note that the images of (6.3) and (6.4) are automatically satisfied. From (6.1) we must have that $X_+Y_+ = ww^{-1} = 1$, and hence that $\phi(X_+)\phi(Y_+) = 1$, which is just (6.19). With this assumption, the images of all the relations in (6.1) are satisfied. This only leaves (6.2), which becomes (6.20).   □

We shall write S_1 for the map $\overline{\phi} \circ S_0 : \mathscr{P}_s \longrightarrow \mathbb{Z}[A,B,d]/K_0$. Note that S_1 is the most general link invariant that can be obtained from the three-variable Kauffman bracket by normalizing using the self-writhe of a diagram. In fact, together with Theorem 13, Theorem 14 can be used to show that the substitution given by (6.16)–(6.18) does not diminish the power of S_0 to distinguish links, as S_0 is determined by S_1.

Theorem 15. *The homomorphism* $\psi : \mathbb{Z}[A,B,d] \longrightarrow \mathbb{Z}[w,w^{-1},\alpha,X_\pm,Y_\pm,x_\pm,y_\pm]$, *defined by* $\psi(A) = x_+$, $\psi(B) = y_+$, *and* $\psi(d) = \alpha$, *induces a homomorphism*

$$\overline{\psi} : \mathbb{Z}[A,B,d]/K_0 \longrightarrow \mathbb{Z}[w,w^{-1},\alpha,X_\pm,Y_\pm,x_\pm,y_\pm]/J_0$$

such that $S_0 = \overline{\psi} \circ S_1$.

Proof. From equations (6.1), (6.3) and (6.4) of Theorem 13, we have the following relations in $\mathbb{Z}[w,w^{-1},\alpha,X_\pm,Y_\pm,x_\pm,y_\pm]/J_0$:

$$\begin{aligned}
w &= X_+ = Y_- = x_+ + \alpha y_+, \\
w^{-1} &= X_- = Y_+ = \alpha x_+ + y_+,
\end{aligned} \tag{6.21}$$

and, since $ww^{-1} = 1$,

$$(x_+ + \alpha y_+)(\alpha x_+ + y_+) = 1. \tag{6.22}$$

Also, from (6.2) and (6.21), we have

$$\alpha = (\alpha x_+ + y_+)y_+ + (x_+ + \alpha y_+)x_+,$$

that is,

$$\alpha = x_+^2 + 2\alpha x_+ y_+ + y_+^2. \tag{6.23}$$

Now equations (6.22) and (6.23) tell us that the homomorphism $\overline{\psi}$ is well defined, as ψ maps the generators of K_0 into J_0. It only remains to show that we can evaluate $S_0(G)$ in terms of the variables x_+, y_+ and α, for all $G \in \mathscr{P}_s$. If so, then the definitions of $\overline{\phi}$ and $\overline{\psi}$ will ensure that $S_0 = \overline{\psi} \circ \overline{\phi} \circ S_0 = \overline{\psi} \circ S_1$. Note that (6.22) implies that $w^{w(G)}$ can always be expressed in terms of x_+, y_+ and α. Thus it only remains to show that $W(G)$ can be expressed in these terms, for all $G \in \mathscr{P}_s$, which we do by induction on the number of edges of G.

Suppose first that G has only bridges and loops. Then $W(G)$ is a product of the variables $X_\pm$, $Y_\pm$ and α. From (6.22), this can be expressed in the required form. Suppose next that G has a positive edge that is neither a bridge nor a loop. Then, from Theorem 6,

$$W(G) = x_+ W(G/e) + y_+ W(G - e).$$

Applying the induction hypothesis to G/e and $G - e$, this can be expressed in terms of x_+, y_+ and α. Otherwise, G has a negative edge e which is neither a bridge nor a loop. Let $\overline{G}$ be the planar dual of G, with all the signs changed. Then $\overline{G}$ and G represent the same link, so $S_0(G) = S_0(\overline{G})$, and $w(G) = w(\overline{G})$, and hence $W(G) = W(\overline{G})$. However, in $\overline{G}$, the edge e is positive, and neither a bridge nor a loop. We thus have that

$$W(G) = W(\overline{G}) = x_+ W(\overline{G}/e) + y_+ W(\overline{G} - e).$$

Since $\overline{G}$ has the same number of edges as G, we can apply the induction hypothesis to $\overline{G}/e$ and $\overline{G} - e$, so $W(G)$ can again be expressed in the required form. This completes the induction, and hence the proof of the theorem. $\qquad\square$

Theorems 13, 14 and 15 enable us to give a simple description of the 'most general' link invariant defined in terms of the coloured Tutte polynomial of the underlying graph.

Theorem 16. *Let K_0 be the ideal of $\mathbb{Z}[A, B, d]$ generated by the relations (6.19) and (6.20), and let $S_1 : \mathscr{P}_s \longrightarrow \mathbb{Z}[A, B, d]/K_0$ be given by evaluating $S(G) = w^{w(G)} W(G)$ under the substitutions*

$$
\begin{aligned}
x_+, y_- &\mapsto A, \\
x_-, y_+ &\mapsto B, \\
w, X_+, Y_- &\mapsto A + dB, \\
w^{-1}, X_-, Y_+ &\mapsto dA + B, \\
\alpha &\mapsto d.
\end{aligned}
$$

Then S_1 is a link invariant, and any link invariant obtained from W, as described at the start of Section 6.2, is of the form $f \circ S_1$, for some ring homomorphism f.

Proof. Theorems 13 and 14 imply that S_1 is a link invariant. Now Theorem 13 implies that any link invariant S obtained from W is of the form $g \circ S_0$, for some ring homomorphism g. However, from Theorem 15, we have that $S_0 = \overline{\psi} \circ S_1$. Hence $S = (g \circ \overline{\psi}) \circ S_1$, completing the proof. $\qquad\square$

Theorem 16 states that the most general link invariant obtained from W is S_1, which is just the three-variable Kauffman bracket $[\cdot](A, B, d)$, multiplied by $(A + dB)^{w(G)}$, evaluated in $\mathbb{Z}[A, B, d]/K_0$, where K_0 is the ideal generated by (6.19) and (6.20). We now turn to more specific examples, in particular the special case of invariants taking values in an integral domain.

6.4. Specific link invariants obtained from the coloured Tutte polynomial

In this section we shall consider special cases of S_1, that is, link invariants of the from $\theta \circ S_1$, where θ is a homomorphism from $\mathbb{Z}[A, B, d]/K_0$ to some other ring R. In the case when R is an integral domain, we can describe all such invariants very easily.

Theorem 17. *Let S be any link invariant obtained from W, as described at the start of Section 6.2, such that S takes values in an integral domain R. Then S can be obtained by composing a ring homomorphism with either the one-variable Jones polynomial, or the invariant $(-1)^{k(L)-1}$, where $k(L)$ is the number of components of a link L.*

Proof. From Theorem 16, it suffices to consider the case $S = \theta \circ S_1$, where θ is a homomorphism from $\mathbb{Z}[A, B, d]/K_0$ to an integral domain R. For notational simplicity, we shall omit θ; thus, for example, we shall write A instead of $\theta(A)$.

Subtracting (6.19) from d times (6.20), we have

$$(d^2 - 1)(AB - 1) = 0.$$

Thus, when our ring R is an integral domain, we must have either $AB = 1$, or $d = \pm 1$.

If $AB = 1$, that is, $B = A^{-1}$, then (6.20) immediately gives $d = -A^2 - A^{-2}$, and, as shown in [13], we obtain the Jones polynomial with the change of variable $t = A^{-4}$.

For the case $d = \pm 1$, we need a little preparation. Let $G \in \mathscr{P}_s$ be any signed plane graph, and $e \in E(G)$ any edge of G. Applying Theorem 6 to all the edges of G other than e, we can express $W(G)$ in the form $fX_\epsilon + gY_\epsilon$, where ϵ is the sign of e, and f and g are polynomials not depending on the sign of e. Thus $W(G) = wf + w^{-1}g$ if e is positive, and $W(G) = w^{-1}f + wg$ if e is negative.

We can now describe $S_1(G)$ in the case that $d = \pm 1$ (strictly speaking, we describe $\theta \circ S_1(G)$ in the case that $\theta(d) = \pm 1$). Suppose first that $d = 1$. Then, from (6.18), $w = A + B$, while $w^{-1} = Y_+ = A + B$. Thus $w = w^{-1}$, and the argument above shows that $W(G)$ is unchanged when the sign of an edge of G is changed. Also, from (6.19) we have that $w^2 = (A + B)^2 = 1$. Since reversing a crossing changes the writhe of a diagram by ± 2, this evaluation of $S_1(G)$ is also invariant under changing the sign of an edge. The same holds if $d = -1$: now $w = A - B = -w^{-1}$, so $W(G)$ changes sign when a crossing is reversed, but $w^2 = -1$, so $w^{w(G)}$ also changes sign, and $S_1(G)$ is unchanged. Now the components of any link can be separated and unknotted if we allow the operation of reversing a crossing. Thus, if G is the graph of any k component link, then $S_1(G) = S_1(E_k) = d^{k-1}$. Since $d = \pm 1$, this completes the proof of the theorem. □

Over more general rings the situation is different. For example, setting $A = 3$, $B = 4$ and $d = -3$ in the ring $\mathbb{Z}/22\mathbb{Z}$ satisfies conditions (6.19) and (6.20), and gives a nontrivial invariant, in that it distinguishes the trefoil from the unknot. At the moment we do not know whether there is a pair of links distinguished by S_1, but not by the Jones polynomial V. Given two different expressions in $\mathbb{Z}[A, B, d]$, to decide whether they are the same requires some calculation, using, for example, the algorithm described in [12]. Having run such calculations using Maple, we know that S_1 does not distinguish the knot 9_{42} from its mirror image, nor 8_8 from 10_{129}. (Our notation for knots is that of [25].)

6.5. Defining a coloured graph polynomial via link diagrams

In [14], Kauffman defines a polynomial $K(G)(A, B, d)$ on signed graphs, *i.e.*, graphs two-coloured with colours $+$ and $-$. This polynomial, which he calls $Q[G](A, B, d)$, is defined by taking the Kauffman bracket defined on link diagrams and extending it to arbitrary signed graphs. In this section we show that proceeding this way around loses generality, i.e., that the polynomial K is a special case of W, from which W cannot be recovered for signed graphs. In fact, we shall show that no polynomial defined on graphs considered as link diagrams can give as much information about the graph as the polynomial W.

Since, in [14], Kauffman shows that $K(G)$ has a spanning tree expansion, we automatically have that $K(G)$ is a special case of W. In particular, from his expansion we see that $K(G)$ is given by $W(G)$ with

$$\begin{aligned} X_+ &= Y_- = A + dB, \\ X_- &= Y_+ = B + dA, \\ x_+ &= y_- = A, \\ x_- &= y_+ = B, \end{aligned}$$

and $\alpha_n = d^{n-1}$. Note that $K(G)$ is in fact a polynomial in two variables, since the sum of the degrees of A and B in any term is just the number of edges of G. It is easily seen that $W(G)$ is more general than $K(G)$, in the following sense.

Theorem 18. *The coloured Tutte polynomial $W(G)$ for signed graphs cannot be recovered from $K(G)$. In fact, setting*

$$K'(G)(A, B, a, b, c, d, e) = a^{a(G)} b^{b(G)} c^{c(G)} e^{d(G)} K(G)(A, B, d),$$

where, as before, $a(G)$ is the number of positive edges of G, $b(G)$ the number of negative edges, $c(G)$ the number of vertices, and $d(G)$ the number of components, there are signed graphs G_1 and G_2 with $K'(G_1) = K'(G_2)$ and $W(G_1) \neq W(G_2)$.

Proof. The only property of K we shall use is the fact that K does not distinguish between a signed plane graph G, and $\overline{G}$, its dual with all signs changed. This can be seen from the above and (3.2), or from the fact that K is defined on link diagrams, as these two graphs correspond to the same link diagram.

Let G_1 be the plane graph with four vertices and six edges, given by a four-cycle, with two consecutive edges doubled. Colour three edges $+$, and three $-$, in an arbitrary way. Now, G_1 and $G_2 = \overline{G}_1$ both have three positive edges, three negative edges, four vertices and one component. As we have $K(G_1) = K(G_2)$, we thus have $K'(G_1) = K'(G_2)$. However, the polynomial W does distinguish G_1 and G_2. In particular, taking $C = 1$, $x_\pm = y_\pm = 1$, $X_\pm = 0$, and $Y_\pm = 2$ (which satisfies the conditions of Theorem 2, as the variables for different colours are the same), we have $W(G_1) = 18$, but $W(G_2) = 14$. $\square$

Note that, in defining the extension K' of K, we did not allow a factor of $w^{w(G)}$. This is because K and K' are defined on arbitrary signed graphs, and it is only signed *plane* graphs that have an associated diagram, and hence a writhe.

Since the proof of Theorem 18 only used the fact that K does not distinguish two signed graphs corresponding to the same link, we have actually shown that $W(G)$ as a signed graph polynomial is more general than any polynomial defined on link diagrams.

References

[1] Alon, N., Frieze, A. and Welsh, D. J. A. (1995) Polynomial time randomized approximation schemes for Tutte–Grothendieck invariants: the dense case. *Random Structures and Algorithms* **6** 459–478.

[2] Annan, J. D. (1995) The complexities of the coefficients of the Tutte polynomial. *Discrete Appl. Math.* **57** 93–103.

[3] Birkhoff, G. D. (1912) A determinant formula for the number of ways of coloring a map. *Ann. Math.* **14** 42–46.

[4] Brylawski, T. and Oxley, J. (1992) The Tutte polynomial and its applications. In *Matroid Applications* (N. White, ed.), Cambridge University Press, pp. 123–225.

[5] Burde, G. and Zieschang, H. (1985) *Knots*, De Gruyter, Berlin.

[6] de la Harpe, P., Kervaire, M. and Weber, C. (1986) On the Jones polynomial. *L'Enseignement Mathématique* **32** 271–335.

[7] Fortuin, C. M. and Kasteleyn, P. W. (1972) On the random-cluster model: I. *Physica* **57** 536–564.

[8] Freyd, P., Yetter, D., Hoste, J., Lickorish, W. B. R., Millett K. and Ocneanu, A. (1985) A new polynomial invariant of knots and links. *Bull. Amer. Math. Soc.* **12** 239–246.

[9] Jaeger, F. (1988) Tutte polynomials and link polynomials. *Proc. Amer. Math. Soc.* **103** 647–654.

[10] Jaeger, F., Vertigan, D. L. and Welsh, D. J. A. (1990) On the computational complexity of the Jones and Tutte polynomials. *Math. Proc. Cam. Phil. Soc.* **108** 35–53.

[11] Jones, V. F. R. (1985) A polynomial invariant of knots via von Neumann algebras. *Bull. Amer. Math. Soc.* **12** 103–111.

[12] Kandri-Rody, A. and Kapur, D. (1984) Algorithms for computing Gröbner bases of polynomial ideals over various Euclidean rings. In *EUROSAM 84 (Cambridge, 1984)*, Vol. 174 of *Lecture Notes in Computer Science*, Springer, Berlin–New York, pp. 195–206.

[13] Kauffman, L. H. (1987) State models and the Jones polynomial. *Topology* **26** 395–407.

[14] Kauffman, L. H. (1989) A Tutte polynomial for signed graphs. *Discrete Appl. Math.* **25** 105–127.

[15] Kauffman, L. H. (1990) An invariant of regular isotopy. *Trans. Amer. Math. Soc.* **318** 417–471.

[16] Lickorish, W. B. R. (1988) Polynomials for links. *Bull. London Math. Soc.* **20** 558–588.

[17] Lipson, A. S. (1988) A note on some link polynomials. *Bull. London Math. Soc.* **20** 532–534.

[18] Lipson, A. S. (1992) Some more states models for link invariants, *Pacific J. Math.* **152** 337–346.

[19] Murasugi, K. (1987) Jones polynomials and classical conjectures in knot theory. *Topology* **26** 187–194.

[20] Murasugi, K. (1989) On invariants of graphs with applications to knot theory. *Trans. Amer. Math. Soc.* **314** 1–49.

[21] Negami, S. (1987) Polynomial invariants of graphs. *Trans. Amer. Math. Soc.* **299** 601–622.

[22] Oxley, J. G. and Welsh, D. J. A. (1979) The Tutte polynomial and percolation. In *Graph Theory and Related Topics* (J. A. Bondy and U. S. R. Murty, eds), Academic Press, New York, pp. 329–339.

[23] Przytycki, J. H. and Traczyk, P. (1987) Invariants of links of Conway type. *Kobe J. Math.* **4** 115–139.

[24] Reidemeister, K. (1948) *Knotentheorie*, Chelsea Publishing Co., New York. Copyright 1932, Julius Springer, Berlin.

[25] Rolfsen, D. (1976) *Knots and Links*, Publish or Perish, Berkeley, 439pp.

[26] Schwärzler, W. and Welsh, D. J. A. (1993) Knots, matroids and the Ising model. *Math. Proc. Cam. Phil. Soc.* **113** 107–139.

[27] Sekine, K., Imai, H. and Tani, S. (1995) Computing the Tutte polynomial of a graph of moderate size. In *Algorithms and Computations*, Vol. 1004 of *Lecture Notes in Computer Science*, Springer, Berlin, pp. 224–233.

[28] Tait, P. G. (1898) On knots I, II, III. In *Scientific Papers, Vol. I*, Cambridge University Press, pp. 273–347.

[29] Thistlethwaite, M. B. (1987) A spanning tree expansion of the Jones polynomial. *Topology* **26** 297–309.

[30] Thistlethwaite, M. B. (1988) Kauffman's polynomial and alternating links. *Topology* **27** 311–318.

[31] Thistlethwaite, M. B. (1988) On the Kauffman polynomial of an adequate link. *Invent. Math.* **93** 285–296.

[32] Traldi, L. (1989) A dichromatic polynomial for weighted graphs and link polynomials. *Proc. Amer. Math. Soc.* **106** 279–286.

[33] Tutte, W. T. (1947) A ring in graph theory. *Proc. Cam. Phil. Soc.* **43** 26–40.

[34] Tutte, W. T. (1954) A contribution to the theory of chromatic polynomials. *Canadian J. Math.* **6** 80–91.

[35] Tutte, W. T. (1967) On dichromatic polynomials. *J. Combin. Theory* **2** 301–320.

[36] Welsh, D. J. A. (1993) *Complexity: Knots, Colourings and Counting*, Vol. 186 of *London Math. Soc. Lecture Notes*, Cambridge University Press.

[37] Welsh, D. J. A. (1995) Randomised approximation schemes for Tutte–Grothendieck invariants, in *Discrete Probability and Algorithms*, Vol. 72 of *IMA Vol. Math. Appl.*, Springer, New York, pp. 133–148.

[38] Whitney, H. (1932) A logical expansion in mathematics. *Bull. Amer. Math. Soc.* **38** 572–579.

[39] Whitney, H. (1932) The coloring of graphs. *Ann. Math.* **33** 688–718.

[40] Zaslavsky, T. (1992) Strong Tutte functions of matroids and graphs. *Trans. Amer. Math. Soc.* **334** 317–347.

Combinatorics, Probability and Computing (1999) 8, 95–107.
© 1999 Cambridge University Press

Notes on Sum-Free and Related Sets

PETER J. CAMERON[1] and PAUL ERDŐS[2]

[1] School of Mathematical Sciences, Queen Mary and Westfield College,
Mile End Road, London E1 4NS, England
(e-mail: p.j.cameron@qmw.ac.uk)

[2] Mathematical Institute of the Hungarian Academy of Sciences,
Reáltanoda u. 13–15, Budapest H-1364, Hungary

Our main topic is the number of subsets of $[1, n]$ which are maximal with respect to some condition such as being sum-free, having no number dividing another, *etc.* We also investigate some related questions.

In our earlier paper [8], we considered conditions on sets of positive integers (sum-freeness, Sidon sequences, *etc.*), and attempted to estimate the number of subsets of $[1, n]$ satisfying each condition, and the Hausdorff dimension of the set of such subsets of the natural numbers. Most of our effort went into sum-free sets, which have been an obsession of the first author for some time [7].

The present paper is a sequel. Our main concern is to provide similar estimates for the maximal subsets of $[1, n]$ or of the natural numbers satisfying such conditions. We also consider some related questions about sum-free sets: what is the largest sum-free subset of $[1, n]$ with given least element; the existence of large sum-free sets containing no long arithmetic progressions; some classes of sum-free sets defined by irrational numbers; and some remarks on random sum-free sets.

1. Sum-free sets

1.1. The number of maximal sum-free sets
A subset S of $\{1, \ldots, n\}$ is *sum-free* if $x, y \in S \Rightarrow x + y \notin S$. (We allow $x = y$ here.) The number $f(n)$ of sum-free sets is known to be $2^{(\frac{1}{2} + o(1))n}$. (This bound was found independently by Alon, Calkin, and Erdős and Granville: see, for example, [3].) We have $f(n) \geqslant c \cdot 2^{n/2}$; we can take any set of odd numbers in $\{1, \ldots, n\}$, or any set of numbers greater than $\frac{1}{2}n$. It is conjectured that $f(n)$ is asymptotically $c \cdot 2^{n/2}$ [8].

Since all the sum-free sets just mentioned lie in just two maximal ones, we might expect that the number $f_{\max}(n)$ of maximal sum-free sets is substantially smaller than $f(n)$. So far we have been unable to find a nontrivial upper bound for $f_{max}(n)$. The following lower bound applies.

Theorem 1. *We have* $f_{max}(n) \geqslant 2^{\lfloor n/4 \rfloor}$.

Proof. Let m be either n or $n-1$, whichever is even. Now let S consist of m together with one of each pair of numbers $\{x, m-x\}$ for odd $x < m/2$. Then S is sum-free. It may not be maximal, but no further odd numbers less than m can be added, so distinct sets S lie in distinct maximal sum-free sets.   □

Problem. Is it true that $f_{max}(n) = o(f(n))$?

We make a few remarks. In order to show that $f_{max}(n) = o(2^{n/2})$, it would suffice to prove such a bound for maximal sum-free sets S satisfying $p = \min(S) = o(n)$ and $\frac{1}{10}n \leqslant |S| \leqslant \frac{2}{5}n$.

For let $p = \min(S)$. Suppose that $n - p < x \leqslant n$. If $x = y + z$ for $y, z \in S$, then $y, z \leqslant n - p$, and $x \notin S$. If not, then x may be added to S without destroying the sum-free property; by maximality, $x \in S$. So S is determined by its intersection with $\{1, \ldots, n - p\}$. Thus the number of maximal sum-free sets with least element p is at most $f(n - p)$, which is $o(2^{n/2})$ if $p > \epsilon n$.

The lower bound on $|S|$ can be assumed because, if $k < \frac{1}{10}n$, then $\binom{n}{k} < 2^{\alpha n}$ with $\alpha < \frac{1}{2}$. The upper bound is deduced from a theorem of Deshouillers, Freiman, Sós and Temkin [9], which asserts that a sum-free set S satisfies one of the following:

(a) S consists of odd numbers
(b) $|S| \leqslant \min(S)$
(c) $|S| \leqslant \frac{2}{5}(n + 2)$.

Clearly there is a unique maximal set satisfying (a). By the previous argument, we may assume that (b) doesn't occur.

The fact that $f_{max}(n) = o(f(n))$ would follow if a conjecture in our earlier paper were true. We assume that n is even in the following discussion; the results are the same (apart from the constants) for odd n.

We showed that the number $f'(n)$ of sum-free sets with least element greater than $n/3$ is asymptotically $c2^{n/2}$ for some constant c (which we estimated). Moreover, if $w(n)$ is any function tending to infinity with n, then almost all of these sets have least element greater than $n/2 - w(n)$. We conjectured that almost all sum-free sets either are of this form, or else consist of odd numbers. (It would follow that $f(n) \sim (c + 1)2^{n/2}$.)

Consider sum-free sets with least element at least $p = n/2 - w + 1$, where $w < n/6$. Such a set S consists of an arbitrary subset T of $[p, n - p]$, an arbitrary subset U of $[n - p + 1, 2p - 1]$, and an arbitrary subset of $[2p, n]$ disjoint from $T + T$. Since there are no restrictions on U, we see that if S is maximal then U is the entire interval $[n - p + 1, 2p - 1]$. So only a proportion $1/2^{3p - n - 1}$ of these sets are maximal. If p is chosen so that $p - n/3 \to \infty$, this proportion is $o(1)$.

Of course, there is only one maximal set of odd numbers.

Another approach involves estimating the number of maximal sets containing a given nonmaximal set. Since each maximal set has cardinality at least $c\sqrt{n}$ (Theorem 2), an affirmative answer to the following question would suffice to give $f_{max}(n) = o(f(n))$.

Problem. Let S be a subset of $[1, n]$ obtained by deleting one term from a maximal sum-free set. Is it true that S lies in only $o(\sqrt{n})$ maximal sum-free sets?

A related question concerns the Hausdorff dimension of the set $\mathscr{S}_{max}$ of maximal sum-free subsets of $\mathbb{N}$. It is known that the set $\mathscr{S}$ of all sum-free sets has dimension $\frac{1}{2}$ (see [8]), so $\dim(\mathscr{S}_{max}) \leqslant \frac{1}{2}$. The $2^{\lfloor n/4 \rfloor}$ maximal sets constructed above are not relevant to this question, since they cannot be prolonged to subsets of $\mathbb{N}$. However, two related constructions give lower bounds.

(a) The sets consisting of 1 and one of each pair $5k + 2$, $5k + 3$ for $k \geqslant 1$ lie in distinct maximal sets; so $\dim(\mathscr{S}_{max}) \geqslant \frac{1}{5}$. (We require 2^k open balls of diameter $2^{-(5k+3)}$ to cover all these sets.)
(b) We do slightly better by taking the sets constructed by including 2 and then odd numbers with gaps of 4 or 6. Again, these lie in distinct maximal sets. The number x_n of restrictions to $[1, 2n]$ satisfies the recurrence $x_{n+3} = x_n + x_{n+1}$. So $x_n \sim c\alpha^n$, where $\alpha^3 = \alpha + 1$, $\alpha > 1$. Thus $\dim(\mathscr{S}_{max}) \geqslant \log \alpha/2 \log 2 = 0.2028 \ldots$.

Before leaving this topic, we note the following.

Theorem 2. *The size $s^*(n)$ of the smallest maximal sum-free subset of $[1, n]$ satisfies $c_1\sqrt{n} \leqslant s^*(n) \leqslant c_2\sqrt{n}$.*

Proof. S is maximal if and only if every element of $[1, n]$ not in S is of the form $x+y$, $x-y$, or $x/2$ for some $x, y \in S$. If $|S| = m$, there are at most $m(m+1)/2 + m(m-1)/2 + m = m(m+1)$ such elements; so $m(m + 2) \geqslant n$, whence $m \geqslant \sqrt{n+1} - 1$.

The upper bound follows from a construction of Hanson and Seyffarth [12]. They gave a set T of $2t + 1$ residues mod $s = t^2 + 5t + 2$ which is sum-free and has the property that any residue not in T is the sum of two members of T (mod s). It is easily seen that the members of these residue classes in $[1, 2s]$ form a maximal sum-free set. $\qquad \square$

1.2. Sum-free sets with given least element

We saw in the preceding section that the minimum element of a sum-free set has a significant effect on the structure. Here we consider the function $g(n, p)$, where $g(n, p)$ is the size of the largest sum-free subset of $\{1, \ldots, n\}$ which has smallest element p.

Theorem 3. *We have*
$$g(n, p) \leqslant \begin{cases} n - p + 1 & \text{if } p > \frac{1}{2}n, \\ p & \text{if } \frac{1}{3}n \leqslant p \leqslant \frac{1}{2}n, \\ \lfloor \frac{1}{2}(n - p) \rfloor + 1 & \text{if } p < \frac{1}{3}n. \end{cases}$$
This bound is attained if either $p \geqslant \frac{1}{5}(n + 2)$ or p is odd.

Proof. If $p > \frac{1}{2}n$, the set $\{p, p+1, \ldots, n\}$ contains $n - p + 1$ integers and is sum-free.

Suppose that $\frac{1}{3}n \leqslant p \leqslant \frac{1}{2}n$. Suppose that S is sum-free, with $\min(S) = p$. For $2p \leqslant x \leqslant n$, S contains at most one of each pair $x - p, x$ of integers; so at least $n - 2p + 1$ integers are excluded, and $|S| \leqslant (n - p + 1) - (n - 2p + 1) = p$. The bound is realized by the set $S = \{p, \ldots, 2p - 1\}$.

Suppose that $p < \frac{1}{3}n$. Let $q = \lfloor (n-p)/2 \rfloor + 1, r = \lfloor (n+p)/2 \rfloor + 1$. Then $p < q$, $p + q = r$, and $q + r - 2 \leqslant n$. Let $k = |S \cap \{p, \ldots, q - 1\}|$ and $l = |S \cap \{q, \ldots, r - 1\}|$. Then at least $k + l - 1$ numbers between $p + q = r$ and n are sums of two members of S and so are excluded; so $|S| \leqslant (n - r + 1) + 1 = \lceil (n - p)/2 \rceil + 1$.

If $n - p$ is even, or if $n - p$ is odd and $n \notin S$, we are finished. If $n - p$ is odd and $n \in S$, then $n - p = (q - 1) + q \notin S$, so we may increase q and r by one in the above argument.

For $\frac{1}{3}(n + 2) \leqslant p \leqslant \frac{1}{3}n$, there is a sum-free set of size $\lfloor \frac{1}{2}(n - p) \rfloor + 1$, namely $\{p, \ldots, q\} \cup \{2q + 1, \ldots, n\}$, where $q = \lfloor \frac{1}{2}(n - p + 1) \rfloor$. For any odd $p \leqslant \frac{1}{3}n$, there is another set of the same size, namely the odd numbers in $\{p, \ldots, n\}$. This completes the proof of the theorem. $\qquad\square$

For even $p \leqslant \frac{1}{5}n$, the results of Deshouillers, Freiman, Sós and Temkin [9] show that $f(n, p) \leqslant \frac{2}{5}(n + 2)$. We do not know if this bound can be attained for $p > 2$. If p is even but not divisible by 5, then the intersection of $\{p, \ldots, n\}$ with either $\{5k + 2, 5k + 3\}$ or $\{5k + 1, 5k + 4\}$ has cardinality roughly $\frac{2}{5}(n - p)$. However, for $p > \frac{1}{6}n$, this is beaten by $\{p, \ldots, 2p - 1\} \cup \{4p - 1, \ldots, 5p - 2\}$, with cardinality $2p$.

1.3. The sets S_α

Let α be any irrational real number, and let

$$S_\alpha = \{k \in \mathbb{N} : \tfrac{1}{3} < \{k\alpha\} < \tfrac{2}{3}\},$$

where $\{x\}$ denotes the fractional part of x. The set S_α is sum-free and has density $\frac{1}{3}$. These sets were used by Erdős [11] (see also Alon and Kleitman [2]) to show that any set S of n integers contains a sum-free subset of size at least $n/3$ (namely, its intersection with some S_α). We mention here the open problem of improving this bound to $n/3 + f(n)$, for any function $f(n)$ that tends to infinity.

Calkin and Erdős [5] showed that S_α is not complete: that is, there are infinitely many numbers not in S_α that have no representation in the form $x + y$ for $x, y \in S_\alpha$. However, S_α is maximal. To see this, let $n \notin S_\alpha$. Suppose that $d = \{n\alpha\} < \frac{1}{3}$ (the other case is similar). Since $\{\{k\alpha\} : k \in \mathbb{N}\}$ is dense in $(0, 1)$, there exists $p \in \mathbb{N}$ such that $\frac{1}{3} < \{p\alpha\} < \frac{2}{3} - d$. Then $\frac{1}{3} + d < \{(p + n)\alpha\} < \frac{2}{3}$; so $p, p + n \in S_\alpha$, and we cannot enlarge S_α to a sum-free set by adding n.

This construction may appear to give a large collection of maximal sum-free sets. However, it follows from the next result that $\dim(\{S_\alpha : \alpha \text{ irrational}\}) = 0$.

Theorem 4. *Let $S_\alpha(n) = S_\alpha \cap \{1, \ldots, n\}$. The number of sum-free sets of the form $S_\alpha(n)$ for fixed n is*

$$1 + \tfrac{1}{2} \sum_{q=1}^{n} \phi(3q) \sim cn^2,$$

where $c = 27/(8\pi^2)$.

Proof. First observe that $S_\alpha = S_{\alpha+m} = S_{m-\alpha}$ for any integer m; so we may assume that $0 < \alpha < \frac{1}{2}$. Now $S_\alpha(n)$ is determined by the least interval containing α whose end-points are $0, \frac{1}{2}$, or of the form $p/(3q)$ where $1 \leq q \leq n$, $1 \leq p < \frac{3}{2}q$, and $3 \nmid p$.

We claim that, if α and β lie in distinct intervals of the above form, then $S_\alpha(n) \neq S_\beta(n)$. So suppose that $S_\alpha(n) = S_\beta(n)$. We show by induction that α and β lie in the same interval $(l/3k, m/3k)$. This is true for $k \leq 2$ by inspection. Suppose that it is true for k but false for $k+1$. Assuming that $\alpha < \beta$, it is clear that this can happen only if

$$\frac{l}{3k} < \alpha < \beta < \frac{l+2}{3k},$$

$$\alpha < \frac{m}{3(k+1)} < \frac{m+1}{3(k+1)} < \beta.$$

These inequalities imply

$$l + \frac{l}{k} < m < l + 1 + \frac{l+2}{k}.$$

Hence $m = l+1$ or $m = l+2$. But, from the form of the intervals, we have $l \equiv 2 \pmod 3$ and $m \equiv 1 \pmod 3$. So $m = l+2 > k$. Then it can be checked that the interval $((l-1)/3(k-2), l/3(k-2))$ contains β but not α: a contradiction.

It follows that the number of sets S_α is one more than the number of rationals $p/(3q)$ satisfying the conditions above. This number is $\sum_{q=1}^{n} \phi(3q)/2$, since half of the integers p with $1 \leq p \leq 3q - 1$ and $(p, 3q) = 1$ are counted.

For the asymptotic estimate, note that $f(n) = \sum_{q=1}^{n} \phi(q)$ is half the number of pairs (p, q) with $1 \leq p, q \leq n$ and $(p, q) = 1$, hence is asymptotically $(3/\pi^2)n^2$. Also, $\phi(3q) = 2\phi(q)$ if $3 \nmid q$, and $\phi(3q) = 3\phi(q)$ if $3|q$. So the sum in question is

$$1 + f(n) + \tfrac{1}{2} \sum_{q=1}^{\lfloor n/3 \rfloor} \phi(q),$$

from which the result follows. $\qquad\square$

Problem. How many of the sets $S_\alpha(n)$ are maximal?

The sets $S_\alpha(n)$ have another interesting property. There is a tendency for large maximal sum-free sets to contain large arithmetic progressions, of size $\Omega(n)$. (Examples include the odd numbers, the set $\{5k+2, 5k+3 : k \in \mathbb{N}\}$, the set $\{\frac{1}{2}n+1,\ldots,n\}$ for even n, and the set $\{p,\ldots,2p-1\} \cup \{4p-1,\ldots,5p-2\}$.) At the Mátraháza meeting, V. Sós asked whether there exist large sum-free sets containing no long arithmetic progression. The largest arithmetic progression contained in $S_\alpha \cap [1, n]$ depends on the approximability of n. The following observation is based on a remark to the authors by I. Ruzsa. Let $\tau = (1 + \sqrt{5})/2$. Recall that S_α has density $1/3$ for any irrational α.

Theorem 5. *The longest arithmetic progression in $S_\tau(n)$ has size $O(\sqrt{n})$.*

Proof. Let $a, a + q, \ldots, a + (m-1)q \in S_\tau(n)$. Then $(m-1)q \leq n$, and $\{a\tau\}, \ldots, \{(a + (m-1)q)\tau\} \in (\frac{1}{3}, \frac{2}{3})$. Hence $\{\pm q\tau\} < 2/(3(m-1))$, and so, for some integer p, we have

$$\left| \tau - \frac{p}{q} \right| < \frac{1}{3(m-1)q}.$$

If $m - 1 > c\sqrt{n}$, then $q < \sqrt{n}/c$, and so $m - 1 > c^2 q$; so $|\tau - p/q| < 1/(3c^2 q^2)$. But, if $c > (5/9)^{1/4}$, then $1/(3c^2) < 1/\sqrt{5}$, and this inequality has only finitely many solutions.   □

However, with random methods, we can do better.

Theorem 6. *There is a function $c(\epsilon)$ such that, for any $\epsilon > 0$, a sum-free set S with density $1/2 - \epsilon$ exists for which $S \cap [1, n]$ contains no arithmetic progression longer than $c(\epsilon) \log n$.*

Proof. First we choose a set T with no long arithmetic progressions. Let $p = 1 - 2\epsilon$, and choose natural numbers independently with probability p. Since the number of m-term arithmetic progressions in $[1, n]$ is at most n^2, the expected number of such progressions in T is at most $n^2 p^m$. If $m > 3 \log n / \log(1/p)$, this expected number is less than $1/n$, and so the probability that such a progression occurs is less than $1/n$. Hence the probability that, for some r, the interval $[1, 2^r]$ contains an arithmetic progression of more than $3r \log 2 / \log(1/p)$ terms is smaller than $\sum 2^{-r} = 1$. So, with positive probability, this never occurs. Let T be such a set. If $T \cap [1, n]$ contains an arithmetic progression of length greater than $3 \log(2n) / \log(1/p)$, then so does $T \cap [1, 2^r]$, where 2^r is the least power of 2 that is not less than n, which is a contradiction since $2^r < 2n$.

To produce a sum-free set, we simply put $S = 2T + 1$.   □

1.4. Random sum-free sets

We next discuss some open problems about random sum-free sets. A random sum-free set S is defined as follows. Consider the natural numbers in turn. If $n = x + y$ with $x, y \in S$, then $n \notin S$; otherwise choose with probability $\frac{1}{2}$ whether $n \in S$ or not (all choices independent).

Let us say that a sum-free set U has *property P* if $\mathrm{Prob}(S \subseteq U) > 0$. Neil Calkin, in his thesis (and see also [4]), showed that property P is equivalent to the condition that, for $n \notin U$, the number of representations $n = x + y$ with $x, y \in U$ grows rapidly (faster than $\log n$) A special case of this, in Cameron [6], is the following. let T be a set of residue classes mod n which is 'complete sum-free', that is, T is sum-free and any residue not in T is the sum of two elements of T (where all calculations are mod n). Examples include the odd numbers ($\{1\}$ mod 2), and the set $\{2, 3\}$ mod 5.

Both these sets can be expressed as S_α for *rational* α. For example, we get the odd numbers for $\alpha = \frac{1}{2}$, and the set $\{5k + 2, 5k + 3 : k \in \mathbb{N}\}$ for $\alpha = \frac{1}{5}$. However, it follows from the result of Calkin and Erdős [5] that S_α does not have property P if α is irrational. Nevertheless, we could ask the following question.

Problem. Is $\mathrm{Prob}(S \subseteq S_\alpha$ for some irrational $\alpha) > 0$?

Analogous constructions of other maximal sum-free sets can be found. The simplest is

$$T_{p,\alpha} = \{k \in \mathbb{N} : \{k\alpha\} \in (p, 2p) \cup (1 - 2p, 1 - p)\},$$

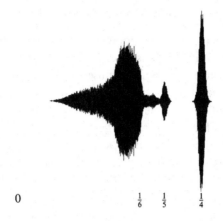

0         $\frac{1}{6}$ $\frac{1}{5}$ $\frac{1}{4}$

Figure 1   The density spectrum.

where $\frac{1}{8} \leqslant p \leqslant \frac{1}{6}$ and α is irrational. This set has density $2p$. It works because, under these conditions on p, the set $U = (p, 2p) \cup (1 - 2p, 1 - p)$ is 'complete sum-free mod 1', that is, it is sum-free and any number not in U is the sum of two numbers in U. Many similar sets can be found.

There are many unsolved problems on random sum-free sets. Here are two.

(a) Does a random sum-free set almost surely have a density?

(b) Is the lower density of a random sum-free set almost surely positive?

Note that an affirmative solution to (b) would imply that a set with property P cannot have density 0.

Empirically, the spectrum of densities seems to be discrete above $\frac{1}{6}$ but to have a continuous part below this value. This could be partly explained if the set of subsets of the sets $T_{p,\alpha}$ above (for all p and α) were shown to have positive measure.

Figure 1 shows an approximation to the density spectrum found by computing many pseudo-random sum-free subsets of $[1, 10000]$.

Another interesting question is the ratio of even to odd members of a random sum-free set. Usually, this ratio appears to be 1 or smaller. (For example, with positive probability, there are no even numbers at all.) However, it tends to a limit greater than 1 with positive probability. For consider the event that all members of S are congruent to a member of T mod $8k + 4$, where

$$T = \{4t + 2 : 0 \leqslant t \leqslant 2k, t \neq k\} \cup \{2k + 1, 6k + 3\}.$$

By the results of Cameron [6], this has positive probability, since the set of residue classes is *complete sum-free* (that is, it is sum-free, and every residue class not in the set is the sum of two classes in the set). Moreover, conditioned on this event, the limiting frequencies of the residue classes are equal; so, the ratio of even to odd numbers tends to k.

This suggests several questions. Let $r(n)$ be the ratio of the number of even numbers in $S \cap [1, n]$ to the number of odd numbers in this set. Does the event $\limsup r(n) = \infty$ have positive probability? Which numbers r have the property that $\limsup r(n) = r$ has positive probability? Does $r(n)$ almost surely tend to a limit?

2. Maximal sets in other families

In our earlier paper, we estimated the number of subsets of $[1, n]$ satisfying various conditions, of which being sum-free was one. In each of these cases, one can ask for similar estimates for the number of maximal subsets satisfying the appropriate condition. We now give results on some of these cases.

Following our earlier notation, we let $\mathscr{A}$ be a class of sets of integers closed under taking subsets; $s(n)$ denotes the size of the largest member of $\mathscr{A}$ contained in $[1, n]$, and $f(n)$ the number of sets in $\mathscr{A}$ contained in $[1, n]$ (so that $2^{s(n)} \leqslant f(n) \leqslant n^{s(n)}$). Also, we let $f_{\max}(n)$ be the number of sets in $\mathscr{A}$ that are maximal with respect to being contained in $[1, n]$. Another interesting parameter is $s^*(n)$, the size of the smallest such maximal set.

We note that the method of choosing a random sum-free set described earlier can be adapted to the situations described below. We hope to investigate this further subsequently.

2.1. A general result

Theorem 7. *Let $\mathscr{A}$ be closed under taking subsets and also closed under translation, and suppose that the size $s(n)$ of the largest member of $\mathscr{A}$ contained in $[1, n]$ satisfies $s(n) = o(n)$. Then $\limsup f_{\max}(n) = \infty$.*

Remark. This theorem applies to the set of *Sidon sequences* (having all pairwise sums distinct), and to the class of sequences containing no k-term arithmetic progression for fixed k. (For Sidon sequences, $s(n) = O(\sqrt{n})$. The fact that $s(n) = o(n)$ in the second case is Szemerédi's theorem [14].)

Proof. In our previous paper, we showed under the same hypotheses that

$$\limsup f(n)/2^{s(n)} = \infty.$$

Now the result follows, since $f(n) \leqslant f_{\max}(n) \cdot 2^{s(n)}$. □

2.2. All partial sums distinct

Let $\mathscr{A}$ be the set of sequences $a_1 < a_2 < \cdots < a_t \leqslant n$ for which all subsums $\sum \epsilon_i a_i$ (with $\epsilon_i = 0$ or 1) are distinct. In [8] we showed that

$$n^{(1+o(1)) \log n / \log 3} \leqslant f(n) \leqslant n^{(1+o(1)) \log n / \log 2}.$$

Theorem 8. *We have $f_{\max}(n) \geqslant f(n)/cn \log n$.*

Proof. Let such a maximal sequence S have t terms. Since the 2^t partial sums are distinct,

$$2^t - 1 \leqslant tn \leqslant cn \log n.$$

But $2^t - 1$ is the number of non-empty subsequences of S. The result follows. □

2.3. Requiring divisibility

Let $\mathcal{A}$ be the family of sequences $a_1 < a_2 < \cdots$ for which a_i divides a_j whenever $i < j$. In [8], we observed that the number $f(n)$ of such sequences in $[1, n]$ satisfies $f(n) \sim cn^\alpha$, where $\alpha = 1.7286\ldots$ is the root of $\zeta(\alpha) = 2$. (ζ is the Riemann zeta-function.)

Maximality of such a sequence means that $a_1 = 1$ and all the factors a_{i+1}/a_i are prime. Hence, the number $g(n)$ of maximal sequences with largest term n is given by the recurrence

$$g(1) = 1, \qquad g(n) = \sum_{\substack{p \text{ prime} \\ p|n}} g(n/p);$$

and we have

$$f_{\max}(n) = \sum_{n/2 < m \leqslant n} g(m).$$

So certainly $f_{\max}(n) \geqslant n/2$.

In fact, $f_{\max}(n)$ grows faster than linearly. For let $n = (2.3.5.7.11.13)^k$. Then $f(n)$ is at least the number of orderings of the factors of n, which is $(6k)!/(k!)^6 > (6^6 - \epsilon)^k$ for large k. Thus,

$$\limsup \log f_{\max}(n)/\log n \geqslant \log(6^6)/\log(2.3.5.7.11.13) = 1.042736\ldots.$$

(Since $\log(p_1 \cdots p_r) \sim \log(r^r)$, where $p_1, \ldots, p_r$ are the first r primes, we cannot do much better by this method.)

Problem. Is it true that $f_{\max}(n) \sim dn^\beta$ for some constants d and β?

2.4. Forbidding divisibility

Let $\mathcal{A}$ be the family of sequences where no term divides another. In our earlier paper, we showed that $c_1^n \leqslant f(n) \leqslant c_2^n$ for large n, where $c_1 = 1.55\ldots$ and $c_2 = 1.59\ldots$. We conjectured, but were unable to prove, that $f(n)^{1/n}$ tends to a limit as $n \to \infty$.

It is clear that $f_{\max}(n) = o(f(n))$. For any maximal subset of $[1, n]$ in which no term divides another must contain all the primes in the interval $(n/2, n]$, so that

$$f_{\max}(n) \leqslant f(n)/2^{\pi(n) - \pi(n/2)}.$$

We cannot yet show that $f_{\max}(n)$ is exponentially smaller than $f(n)$. Some remarks on this follow.

First, the maximum size $s(n)$ of such a set is $\lceil n/2 \rceil$. For, given any odd number $m \leqslant n$, at most one number of the form $2^a \cdot m$ can belong to the set. If the bound is attained, then the set contains one number of this form for each value of a. Such a set is described by a function $a(m)$ from the odd numbers in $[1, n]$ to the nonnegative integers, satisfying

(a) $a(m) \leqslant \log(n/m)/\log 2$

(b) if m_1 divides m_2, then $a(m_1) > a(m_2)$.

The number of functions satisfying (a) is

$$\prod r^{\lceil n/2^{r+1}\rceil} \sim c^n,$$

where $c = \prod r^{1/2^{r+1}} = 1.2891\ldots$. This is exponentially smaller than our lower bound for $f(n)$.

There is also an exponential lower bound for the number of such subsets of cardinality $\lceil n/2 \rceil$. For each integer x in $(n/3, n/2]$, include either x or $2x$ in the set; then include all integers in $(n/2, 2n/3]$, and all odd integers in $(2n/3, n]$. This gives about c^n sets, where $c = 2^{1/6} = 1.12246\ldots$.

We also have some information on the small maximal sets. To avoid trivialities, we do not allow the number 1.

Theorem 9. *The size $s^*(n)$ of the smallest maximal set in $\mathscr{A}$ contained in $[2, n]$ is $\pi(n)$. There are at least $2^{cn/\log n}$ maximal sets of this size.*

Proof. Let $a_1 < a_2 < \cdots < a_k \leqslant n$ be such a maximal set. We prove that, given any set P of m primes in $[1, n]$, there are at least m terms of the sequence divisible by some prime in P. The proof is by induction on m. For $m = 1$, if there were a prime dividing no term of the sequence, it could be added to the sequence.

Suppose that the result is true for $m - 1$ but false for m. Let P be a set of m primes whose members divide fewer than m terms of the sequence. Let p be the smallest prime in P, and p^k the largest power of p dividing any term in the sequence. It cannot occur that a term in the sequence is divisible by p^k and no other prime in P, else the remaining $m - 1$ primes would divide fewer than $m - 1$ terms. So there is some term $p^k q x$ in the sequence, where q is a prime greater than p. Then $p^{k+1} < p^k q x \leqslant n$, and we may add p^{k+1} to the sequence. This completes the induction.

The proof shows that the conditions of Hall's theorem hold; so, if $p_1 < p_2 < \cdots < p_r$ are the primes in $[1, n]$, with $r = \pi(n)$, then we can number the sequence as $a_1, a_2, \ldots, a_r$ so that p_i divides a_i for all i. The argument shows that if a set of m primes divide only m terms of the sequence, and p_i is the smallest of the primes, then a_i is divisible by no other prime in the set. Applying this inductively to $p_i, p_{i+1}, \ldots, p_r$, we see that, if p_j divides a_i, then $j \leqslant i$.

We construct many maximal sequences as follows. Start with the set of primes in $[1, n]$. Let p be any prime less than $\sqrt{n}$. For each prime q satisfying $\sqrt{n} < q \leqslant n/p$, multiply q by any power of p (so that the result is still less than n). Then replace p by any power of itself strictly greater than the powers used in the first part of the construction. The result is still in $\mathscr{A}$ and is still maximal.

If we choose $p = 2$, and replace some primes $q \in (\sqrt{n}, n/2]$ by $2q$ and then replace 2 by 4, we obtain $2^{\pi(n/2) - \pi(\sqrt{n})}$ maximal sets of size $\pi(n)$. This is of order $2^{cn/\log n}$ as claimed: the more general construction merely improves the value of c slightly. $\square$

Problems. How many maximal sets are there? Does $f_{\max}(n)^{1/n}$ tend to a limit? How many maximal sets of size $\pi(n)$?

2.5. Any two terms coprime

Let $\mathscr{A}$ be the set of sequences with any two terms coprime. Apart from 1, the terms of such a sequence have disjoint sets of prime factors, so there are at most $\pi(n)$ of them. We proved in [8] that the number of sequences is at least $e^{(1/2+o(1))\sqrt{n}}2^{\pi(n)}$ sequences; so there are at least $e^{(1/2+o(1))\sqrt{n}}$ maximal ones.

Also, there are at most $e^{(2+o(1))\sqrt{n}}$ maximal sequences. For such a sequence contains at most $t = \pi(\sqrt{n})$ nonprime terms, which can be chosen in at most

$$\sum_{i=0}^{t} \binom{n}{i} = e^{(2+o(1))\sqrt{n}}$$

ways; then we must include all primes not dividing these chosen terms.

2.6. No two terms coprime

Let $\mathscr{A}$ be the set of sequences of integers with no two terms coprime. We proved in [8] that $f(n) \leqslant n \cdot 2^{\lfloor n/2 \rfloor}$. On the other hand, $f(n) \geqslant 2^{\lfloor n/2 \rfloor}$, since any set of even numbers belongs to $\mathscr{A}$. This construction gives only one maximal sequence. However, we show that there are relatively many of them.

Theorem 10. *We have* $f_{\max}(n) \geqslant 2^{n^{c/\log\log n}}$.

We deduce the bound from the following construction. An *antichain* in $\mathscr{P}[1,m]$ is a family of subsets of $[1,m]$ no one of which contains another. The *dual*, or *blocker*, of the antichain $\mathscr{X}$ is the antichain $\mathscr{X}^*$ consisting of all sets that are minimal with respect to intersecting every set in $\mathscr{X}$. It is known that $\mathscr{X}^{**} = \mathscr{X}$ (Edmonds and Fulkerson [10]; see Seymour [13] for a short proof). We call $\mathscr{X}$ *self-dual* if $\mathscr{X}^* = \mathscr{X}$.

Theorem 11. *There are at least*

$$2^{\binom{2k-1}{k-1}-1} \Big/ \binom{2k-1}{k-1}^{\binom{2k}{k-1}/2^k} = 2^{(2-o(1))^m}$$

self-dual antichains in $\mathscr{P}([1,m])$, *for* $m = 2k$.

Proof. We call an antichain $\mathscr{X}$ *sub-dual* if $\mathscr{X} \subseteq \mathscr{X}^*$. Now $\mathscr{X}$ is sub-dual if and only if it has the following properties:

(a) $\mathscr{X}$ is intersecting;
(b) for any $X \in \mathscr{X}$ and $x \in X$, there exists $Y \in \mathscr{X}$ such that $X \cap Y = \{x\}$.

If this holds, and Y is any member of $\mathscr{X}^* \setminus \mathscr{X}$, then for each $y \in Y$ there exists $X \in \mathscr{X}$ with $X \cap Y = \{y\}$ (else $Y \setminus \{y\}$ would meet every set in $\mathscr{X}$, and Y would not be minimal). Hence $\mathscr{X} \cup \{Y\}$ is also sub-dual. It follows that any sub-dual antichain can be enlarged to a self-dual antichain.

Our strategy is as follows. If $m = 2k$, any choice of one from each complementary pair of k-subsets of $[1,m]$ gives an intersecting antichain. We show that, for many choices, condition (b) is also satisfied, so that we have many sub-dual antichains. These can be enlarged to self-dual antichains; but in this enlargement process, no further k-sets will be added, so that the resulting self-dual antichains are all distinct.

A choice C of one of each complementary pair of k-sets can be regarded as a vertex of the hypercube of dimension $\binom{2k}{k}/2 = \binom{2k-1}{k-1} = N$, say. So there are are 2^N choices altogether. Now the failure of condition (b) means that there is a $(k-1)$-subset A such that every k-set containing it is in C. (Take $A = X \setminus \{x\}$, with x and X witnessing a failure of (b).) Call such a subset *bad*. Now, in a random choice, the probability that a given set A is bad is $1/2^{k+1}$; so the expected number of bad sets is $\binom{2k}{k-1}/2^{k+1} = \mu$, say. It follows that, for at least half of all possible choices, the number of bad sets is not greater than 2μ.

Suppose that A is bad (for a choice C), so that every k-set containing A is in C. Then any $(k-1)$-set disjoint from A lies in at least two k-sets not in C. Hence, if we replace any k-set containing A by its complement, we obtain a new choice C' in which A is no longer bad, and no new bad set is created. Note that C and C' are adjacent in the N-dimensional hypercube. It follows that a choice with r bad sets lies at distance r in the hypercube from a choice with no bad sets (a *good choice*, we will say). So the balls of radius 2μ centred on the good choices cover at least half the vertices of the hypercube.

The size of such a ball is not greater than $N^{2\mu}$. So there are at least $2^{N-1}/N^{2\mu}$ good choices, as claimed. $\qquad\square$

Proof of Theorem 10. Choose m such that $p_1 p_1 \cdots p_m \leqslant n$, where $p_1, p_2, \ldots, p_m$ are the first m primes. Let $\mathscr{X}$ be a self-dual antichain on $[1, m]$. For each $X \in \mathscr{X}$, let

$$p(X) = \prod_{i \in X} p_i.$$

Let A be the set of numbers in $[1, n]$ divisible by $p(X)$ for some $X \in \mathscr{X}$. Since $\mathscr{X}$ is intersecting, no two members of A are coprime. If r has a common factor with each number in A, then the set $Y = \{i \in [1, m] : p_i | r\}$ meets every set in $\mathscr{X}$. By the self-duality, Y contains some set $X \in \mathscr{X}$, whence $p(X)|r$, and $r \in A$. So A is maximal.

Now $m^m \sim n$, so $m \geqslant c \log n / \log \log n$. Thus the number of maximal sets in $\mathscr{A}$ is at least

$$2^{(2-o(1))^m} \geqslant 2^{n^{c/\log\log n}}.$$

$\qquad\square$

Acknowledgements

Some of this work was done at the DIMANET workshop held in Mátraháza, Hungary, in October 1995. We thank the organizers for their support, and the participants for their helpful comments.

The work was completed during the second author's visit to London in summer 1996, shortly before his death. The first author regrets that further joint work promised in the paper will not now be carried out.

References

[1]   Ahlswede R. and Khachatrian, L. H. (1994) On extremal sets without coprimes. *Acta Arith.* **66** 89–99.

[2]   Alon, N. and Kleitman, D. J. (1990) Sum-free subsets. In *A tribute to Paul Erdős* (A. Baker, B. Bollobás and A. Hajnal, eds), Cambridge University Press, Cambridge, pp. 13–26.

[3]   Calkin, N. J. (1990) On the number of sum-free sets. *Bull. London Math. Soc.* **22** 141–144.

[4]   Calkin, N. J. On the structure of a random sum-free set. To appear.

[5]   Calkin, N. J. and Erdős, P. (1996) On a class of aperiodic sets. *Math. Proc. Cambridge Philos. Soc.* **120** 1–5.

[6]   Cameron, P. J. (1987) On the structure of a random sum-free set. *Probab. Theory Rel. Fields* **76** 523–531.

[7]   Cameron, P. J. (1987) Portrait of a typical sum-free set. In *Surveys in Combinatorics 1987* (C. Whitehead, ed.), Vol. 123 of *London Math. Soc. Lecture Notes*, Cambridge University Press, Cambridge, pp. 13–42.

[8]   Cameron, P. J. and Erdős, P. (1990) On the number of sets of integers with various properties. In *Number Theory* (R. A. Mollin, ed.), Walter de Gruyter, Berlin, pp. 61–79.

[9]   Deshouillers, J.-H., Freiman, G., Sós, V. and Temkin, M. On the structure of sum-free sets. To appear.

[10] Edmonds, J. and Fulkerson, D. R. (1970) Bottleneck extrema. *J. Combin. Theory Ser. A* **8** 299–306.

[11] Erdős, P. (1965) Extremal problems in number theory. *Proc. Symp. Pure Math.* **8** 181–189.

[12] Hanson, D. and Seyffarth, K. (1984) *K*-saturated graphs of prescribed minimal degree. *Congressus Numerantium* **42** 169–182.

[13] Seymour, P. D. (1976) The forbidden minors of binary clutters. *J. London Math. Soc.* **12** 356–360.

[14] Szemerédi, E. (1975) On sets of integers containing no *k* elements in arithmetic progression. *Acta Arith.* **27** 299–345.

Combinatorics, Probability and Computing (1999) **8**, 109–129.

Geometrical Bijections in Discrete Lattices

HANS-GEORG CARSTENS, WALTER A. DEUBER,
WOLFGANG THUMSER and ELKE KOPPENRADE

Universität Bielefeld, Postfach 10 01 31, D-33501 Bielefeld, Germany
(e-mail: deuber@mathematik.uni-bielefeld.de)

We define *uniformly spread sets* as point sets in d-dimensional Euclidean space that are wobbling equivalent to the standard lattice $\mathbb{Z}^d$. A linear image $\varphi(\mathbb{Z}^d)$ of $\mathbb{Z}^d$ is shown to be uniformly spread if and only if $\det(\varphi) = 1$. Explicit geometrical and number-theoretical constructions are given. In 2-dimensional Euclidean space we obtain bounds for the wobbling distance for rotations, shearings and stretchings that are close to optimal. Our methods also allow us to analyse the discrepancy of certain billiards. Finally, we take a look at paradoxical situations and exhibit recursive point sets that are wobbling equivalent, but not recursively so.

1. Introduction

When solving Tarski's circle squaring problem, Laczkovich introduced the notion of a 'uniformly spread set'. Taking a closer look at the combinatorial ideas behind his approach, we consider mappings with a 'wobbling property'. A mapping $f : X \to Y$ (X, Y subsets of a metric space) is a wobbling mapping if no point in X is moved too far: formally, if

$$\sup_{x \in X} d(x, f(x)) < \infty.$$

Wobbling mappings occur in many real-world situations: rounding in numerical analysis, image processing, distortion of crystals, and earthquakes are typical examples. A more sophisticated terminology would call wobbling bijections 'bounded variations of the identity' [4].

We call subsets of a metric space 'equivalent' if there exists a wobbling bijection between them. In particular, uniformly spread sets are those which are equivalent to a lattice $\mathbb{Z}^n$ considered as Euclidean space.

The paper is organized as follows. After a short technical introduction we give some general useful information. Continued fractions will play a crucial role in the effective construction of certain wobbling bijections.

In Section 4 we give a constructive proof for the fact that any linear mapping $f : \mathbb{R}^n \to$

$\mathbb{R}^n$ with determinant 1 transforms the lattice $\mathbb{Z}^n$ into a uniformly spread set. It suffices to consider the Jordan normal form with the typical matrices

$$\begin{pmatrix} 1 & \lambda \\ 0 & 1 \end{pmatrix} \qquad \begin{pmatrix} \cos\alpha & -\sin\alpha \\ \sin\alpha & \cos\alpha \end{pmatrix} \qquad \begin{pmatrix} a & 0 \\ 0 & 1/a \end{pmatrix}$$

$$\text{shearing,} \qquad\qquad \text{rotation,} \qquad\qquad \text{rectangular lattice.}$$

In fact shearings will play an important role for an engineering approach to such problems leading to computer visualization. In particular, this method already yields a uniform upper bound in terms of a matrix norm (Theorem 4.1) for the wobbling distances, which was not initially obvious.

Also, by means of continued fractions and the Sturmian sequence from billiards theory, we will indicate another constructive method and explain this in particular for rectangular lattices. As a side result we shall investigate the discrepancy of a billiards problem. We shall conclude the paper with results on recursive wobblings and an outlook on paradoxical situations [5].

2. General definitions

By $\mathbb{N} = \{1, 2, 3, \ldots\}$ we denote the natural numbers. $\mathbb{N}_0$ stands for the set of nonnegative integers and $\mathbb{R}^+$ for the set of nonnegative reals. For a rational or real $x \in \mathbb{R}$, let $[x]$ be the nearest integer, where ties are broken by rounding to the smaller integer. Let (M, d) be a metric space and $X, Y \subseteq M$. A mapping $f : X \to Y$ is a wobbling mapping if $\sup_{x \in X} d(x, f(x)) < \infty$ and the supremum is called the wobbling distance of f. The composition of two wobbling mappings has the same quality, and this also holds for the inverse of a wobbling bijection. X and Y are called 'wobbling equivalent' or 'equivalent' $(X \equiv Y)$ if there exists a wobbling bijection from X onto Y. The classical Cantor–Bernstein argument shows that X and Y are equivalent if and only if there exist wobbling injections $f : X \to Y$ and $g : Y \to X$. We shall consider lattices in the Euclidean plane with the metric given by the Euclidean norm $\|\cdot\|$. For discrete metric spaces one can use a graph-theoretical approach to wobbling mappings using the Hall–Rado theorem, as follows.

For $k \in \mathbb{R}^+$ and $X, Y \subseteq M$, let $C_k(X, Y)$ be the bipartite k-neighbourhood graph with parts X, Y and edges defined by $d(x, y) \leqslant k$. Let X, Y be subsets of a discrete metric space. Then the following statements are equivalent.

(i) There exists an injective wobbling mapping $f : X \to Y$.

(ii) There exists k, such that $C_k(X, Y)$ satisfies the Hall condition: for every finite subset X' of X the cardinality $|N_k(x') \cap Y|$ of the k-neighbourhood of X' in Y satisfies $|N_k(x) \cap Y| \geqslant |X'|$.

The continued fraction expansion $[a_0, a_1, \ldots]$, where $a_i \in \mathbb{N}_0$, of a real $1/\alpha > 0$ is defined by

$$1/\alpha = a_0 + \cfrac{1}{a_1 + \cfrac{1}{a_2 + \cdots}}.$$

Conversely, it is easy to exhibit the continued fraction expansion of any $1/\alpha$ by the generalized Euclidean algorithm for the pair $(1, \alpha)$. Let

$$
\begin{aligned}
1 &= \alpha a_0 + r_0, \\
\alpha &= r_0 a_1 + r_1, \\
r_0 &= r_1 a_2 + r_2, \\
r_i &= r_{i+1} a_{i+2} + r_{i+2} \quad \text{for } i \in \mathbb{N}.
\end{aligned} \tag{2.1}
$$

Then $1/\alpha = [a_0, a_1, \ldots]$, and if $\alpha > 1$ we have $a_0 = 0$ and $a_i \in \mathbb{N}$ for $i \geqslant 1$, yielding $\alpha = [a_1, a_2, a_3, \ldots]$.

This definition possibly differs from standard notation by a shift, as we start the Euclidean algorithm with $(1, \alpha)$ instead of $(\alpha, 1)$. This seems to be more convenient for our purposes.

3. Uniformly spread sets

In his epochal paper [8], Laczkovich introduced the notion of uniformly spread sets and used it for solving the Tarski circle squaring problem. He immediately established the connection between uniformly spread sets, discrepancies and measure theory. The crucial step was the 2-dimensional case. The generalization to $\mathbb{R}^d$ then relies on some technical details. For simplicity's sake we only consider wobbling equivalences to the standard lattice $\mathbb{Z}^d$. The more general case of $(a\mathbb{Z})^d$ just needs some rescaling.

Definition 3.1. $X \subseteq \mathbb{R}^d$ is uniformly spread *if X is wobbling equivalent to Z^d.*

Laczkovich's idea may most easily be explained in dimension 2. Let $X \subset \mathbb{R}^2$. In order to test whether X is uniformly spread, choose a collection of 'well-behaved' test regions C with boundary ∂C.

Then the discrepancy $|X \cap C| - |\mathbb{Z}^2 \cap C|$ should be small for all possible test sets C, for otherwise no wobbling bijection $X \leftrightarrow \mathbb{Z}^2$ can exist. Now note that a wobbling bijection cannot map points across the boundary ∂C, which are too far away from ∂C, implying that the above discrepancy should be small compared to the boundary as measured by a suitable measure.

$|\mathbb{Z}^2 \cap C|$ is approximately the 'area' $\mu_2(C)$ of C. Thus, hopefully, the discrepancy $\delta_C(X, \mathbb{Z}^2) = |X \cap C| - \mu_2(C)|$ is small compared to the boundary. For measuring the latter, Laczkovich suggests various possibilities.

- Let $\mu_1(\partial C)$ be the perimeter of C measured in the 1-dimensional measure.
- Let $\mu_2(N_1(\partial C))$ be the area of the 1-neighbourhood of $\partial(C)$.

Theorem 3.1 (Laczkovich). *For every discrete set $X \subset R^d$, the following statements are equivalent.*

(i) *There exists a positive constant K such that*

$$
\delta_C(X, \mathbb{Z}^d) \leqslant K \mu_d(N_1(\partial C))
$$

for every bounded Hausdorff measurable set $C \subset \mathbb{R}^d$.

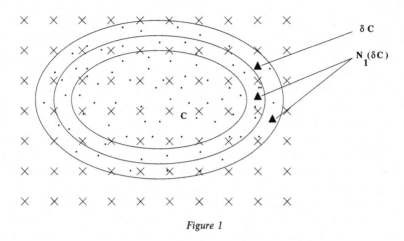

Figure 1

(ii) *There exists a positive constant K such that*

$$\delta_C(X, \mathbb{Z}^d) \leqslant K \circ \mu_{d-1}(\partial C)$$

for every C that is a finite union of unit cubes.

(iii) *X is uniformly spread.*

4. Geometrical constructions of wobbling equivalences in Euclidean lattices

Laczkovich's theorem gives a marvellous abstract characterization for uniformly spread sets. Nevertheless, it depends on the Rado–Hall theorem, which invokes a compactness argument. Therefore it does not give explicit wobbling bijections nor reasonably good upper bounds for the wobbling distance. Starting with the 2-dimensional case, we shall provide such bijections for elementary geometric situations and will finally indicate the general Euclidean case.

Let τ be a translation of the real line. Then $\mathbb{Z}$ and $\tau\mathbb{Z}$ are wobbling equivalent with wobbling distance $\leqslant 1/2$. It suffices to consider $\varphi(n) = n + \tau - [\tau]$ to see that $\tau(\mathbb{Z}) \equiv \mathbb{Z}$ and the wobbling distance is $\leqslant 1/2$.

The next interesting case is the shearing with respect to an arbitrary basis b_1, b_2; $S = \left(\begin{smallmatrix} 1 & \lambda \\ 0 & 1 \end{smallmatrix}\right)$. Each line $g_a = tb_1 + ab_2$ ($a \in \mathbb{Z}$, fixed) parallel to b_1 is mapped to itself and the restriction of S to g_a is a translation with translation vector $\lambda a b_1$, which of course depends on a. Nothing prevents us to define the wobbling bijection on each line g_a separately by $\varphi(tb_1 + ab_2) = (t + \lambda a - [\lambda a])b_1 + ab_2$. We leave it to the reader to verify that such a φ is bijective when considered on the lattice generated by b_1, b_2. The wobbling distance is as follows. Let $x = x_1 b_1 + x_2 b_2$, where $x_1, x_2 \in \mathbb{Z}$. Then $\|\varphi(x) - x\| = \|(\lambda x_2 - [\lambda x_2])b_1\| \leqslant 1/2\|b_1\|$, which is independent of x.

The next problem is to show that any two 2-dimensional lattices with unit fundamental domain are wobbling equivalent. It suffices to observe that the transformation may be obtained as a product of at most three shearings. To see this, take the basis b_1, b_2 for the first lattice and c_1, c_2 for the second lattice, oriented in the same way.

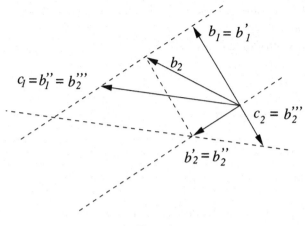

- For the first shearing take the fix vector b_1 and shear parallel to it in such a way that b_2 becomes b'_2 parallel to the line b_1, c_1.
- For the second shearing take b'_2 as axis and shear parallel to it in such a way that b_1 is mapped to c_1. Note that this is possible by construction of b'_2 in the first shearing.
- Finally, for the third shearing keep c_1 fixed and move b'_2 into c_2. This is possible as all fundamental domains have unit area rendering the line b'_2, c_2 parallel to c_1.

As before, it is easy to give an upper bound for the wobbling distance of the product of the three shearings by

$$\|\varphi(x) - x\| \leqslant 1/2 \|b_1 \pm b'_2 \pm c_1\|. \tag{4.1}$$

For the upper bound one has to take the worst combination of $\pm$ signs. In order to evaluate this we take b_1, b_2 to be the orthonormal basis in $\mathbb{R}^2$ and express all vectors in this basis, obtaining

$$\|\varphi(x) - x\| \leqslant \frac{1}{2}\left(1 + \sqrt{1 + \left(\frac{c_{11} - 1}{c_{12}}\right)^2} + \|c_1\|\right) \tag{4.2}$$

$$\leqslant \frac{1}{2}\left(1 + \left(1 + \frac{1}{|c_{12}|}\right)\sqrt{(|c_{11}| + 1)^2 + c_{12}^2}\right), \tag{4.3}$$

where $c_1 = c_{11}b_1 + c_{12}b_2$.

Note that we required $c_{12} \neq 0$ for these calculations. Without loss of generality this may always be achieved by elementary transformations leaving the lattices as point sets unchanged. The transformation mapping b_1, b_2 to $b_2, -b_1$ exchanges the columns in the matrix representation of c_1, c_2, and analogously rows are exchanged by transforming the basis c_1, c_2 to $c_2, -c_1$. As both lattices have unit cells we may assume that at least one coefficient, say c_{12}, does not vanish. We even can do a little better than that.

Fact. *Let b_1, b_2 be the standard basis in $\mathbb{R}^2$ and c_1, c_2 be independent vectors with $\det(c_1, c_2) = \pm 1$. In the representation $c_1 = c_{11}b_1 + c_{12}b_2, c_2 = c_{21}b_1 + c_{22}b_2$ at least one of the coefficients satisfies $|c_{ij}| \geqslant \sqrt{2}/2$.*

Proof. If not, we obtain

$$| \det(c_1 c_2)| \leqslant |c_{11}| |c_{22}| + |c_{12}| |c_{21}| < \frac{\sqrt{2}}{2} \frac{\sqrt{2}}{2} + \frac{\sqrt{2}}{2} \frac{\sqrt{2}}{2} = 1. \tag{4.4}$$

$\square$

Combining this fact with (4.3), we obtain the following.

Theorem 4.1. *Let b_1, b_2 be the standard orthonormal basis in $\mathbb{R}^2$ and c_1, c_2 be a basis with unit fundamental domain. Then the generated lattices are wobbling equivalent and the wobbling distance is at most*

$$5 \max |c_{ij}|. \tag{4.5}$$

$\square$

It is remarkable that one can have such a simple universal upper bound for the wobbling distance using only a constant multiple of the max norm. This was not evident to everyone from the beginning.

Remark. By induction one verifies that a corresponding theorem holds in higher dimensions, too.

By carefully inspecting particular cases such as rotated and rectangular lattices we can obtain even better results.

4.1. Rectangular lattices

$$L_a : (c_1, c_2) = \begin{pmatrix} a & 0 \\ 0 & 1/a \end{pmatrix}, \quad a \in \mathbb{R}, \ |a| > 1 \tag{4.6}$$

Theorem 4.2. *Let L_1 be the standard lattice and L_a be a rectangular lattice. Then L_1 and L_a are wobbling equivalent and the wobbling distance is at most $\frac{3+a}{2}$.* $\square$

Note that the obvious lower estimate for the wobbling distance is $\frac{1+a}{2}$ for irrational a. Moreover, Sudmeier [10] showed that, by using a very involved construction with continued fractions, the upper bound can be improved to $\frac{1+a}{2}$, which he shows to be optimal for irrational α. Anyway, asymptotically our result is close to optimal.

Proof. Because of $c_{12} = 0$ and $c_{21} = 0$ in (4.6), we first take the reflected lattice generated by

$$\begin{pmatrix} 0 & -a \\ 1/a & 0 \end{pmatrix},$$

obtaining $c_{12} = \frac{1}{a}$. Using (4.2), obtain

$$\|\varphi(x) - x\| = \frac{1}{2}\left(1 + \sqrt{1 + a^2} + \frac{1}{a}\right) \leqslant 1 + \frac{\sqrt{1 + a^2}}{2} \leqslant \frac{3 + a}{2}. \tag{4.7}$$

$\square$

Our approach does not differentiate between rational and irrational *as*, and gives a universal linear upper bound.

4.2. Rotated lattices

$$L_\alpha : (c_1, c_2) = \begin{pmatrix} \cos \alpha & -\sin \alpha \\ \sin \alpha & \cos \alpha \end{pmatrix}. \tag{4.8}$$

By symmetry one may assume $0 \leqslant \alpha \leqslant \frac{\pi}{4}$.

Theorem 4.3. *The standard lattice $\mathbf{Z}^2$ and the rotated lattice L_α are equivalent and the wobbling distance is at most $\frac{\sqrt{5}}{2}$.*

Proof. Using (4.1), obtain with $t = \tan \frac{\alpha}{2}$

$$\|\varphi(x) - x\| \leqslant \frac{1}{2}\left\|\begin{pmatrix} 1 \\ 0 \end{pmatrix} \pm \begin{pmatrix} \frac{1 - \cos \alpha}{\sin \alpha} \\ -1 \end{pmatrix} \pm \begin{pmatrix} \cos \alpha \\ \sin \alpha \end{pmatrix}\right\|$$

$$\leqslant \frac{1}{2}\left\|\begin{pmatrix} 1 \pm t \pm \frac{1 - t^2}{1 + t^2} \\ \mp 1 \pm \frac{2t}{1 + t^2} \end{pmatrix}\right\|$$

$$\leqslant \frac{1}{2}\sqrt{3 + 2\left(\frac{1 - t^2}{1 + t^2}\right) + t^2} \leqslant \frac{\sqrt{5}}{2}. \tag{4.9}$$

$\square$

For irrational trigonometric functions of α, the lower bound for the wobbling distance is $\sqrt{2}/2$, as infinitely often the vertices of the rotated lattice L_α are arbitrarily close to midpoints of the squares in the standard lattice. Intuitively, the upper bound $\frac{\sqrt{5}}{2} = 1.1180\ldots$ seems to be optimal for certain angles: at least for some small angles one may expect that infinitely often the vertices of the rotated lattices are arbitrarily close to midpoints of the edges of the standard lattice. Intuitively, the rounding along the edges cannot go on indefinitely. Finally one has to jump diagonally with distance close to $\frac{\sqrt{5}}{2}$.

We checked this intuition numerically for $\alpha = 2.2°$. The computer showed, by the aid of a fast version of Hall's algorithm applied to the neighbourhood graph $G_{1.1170}(L_1, L_\alpha)$, that there is no perfect matching covering the $10^3 \times 10^3$ square centred at the origin. This shows that the optimal wobbling distance for $\alpha = 2.2°$ is between 1.1170 and 1.1180.... We did not care for better bounds.

Remark. As rotations are orthogonal mappings, one immediately hopes that any rotation is a three-fold product of shearings along the standard orthogonal basis. Trying

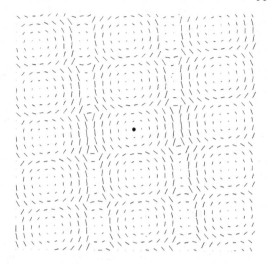

Figure 3

to write

$$\begin{pmatrix} \cos\alpha & -\sin\alpha \\ \sin\alpha & \cos\alpha \end{pmatrix} = \begin{pmatrix} 1 & \lambda_1 \\ 0 & 1 \end{pmatrix} \begin{pmatrix} 1 & 0 \\ \lambda_2 & 1 \end{pmatrix} \begin{pmatrix} 1 & \lambda_3 \\ 0 & 1 \end{pmatrix} = S_3 \circ S_2 \circ S_1 \qquad (4.10)$$

with unknown $\lambda_1, \lambda_2, \lambda_2$, one immediately obtains for $t = \tan\alpha/2$ that

$$\lambda_1 = \lambda_3 = -t \quad \text{and} \quad \lambda_2 = \sin\alpha. \qquad (4.11)$$

For a matrix A one may introduce the notation $[A]$ for the mapping taking the value $[A]x$ as the vector obtained from Ax by coordinatewise rounding to the nearest integer. Observe that $[A]$ in general is not continuous, and mathematically nasty. However, $[A]$ is easily handled by computers. It is certainly not compatible with matrix composition. We use the natural composition as iterated mappings. One obtains the following wobbling bijection between the standard lattice and the rotated lattice L_α:

$$\varphi = S_3 \circ S_2 \circ S_1 \circ [S_1^{-1}] \circ [S_2^{-1}] \circ [S_3^{-1}], \qquad (4.12)$$

which is easily implemented and may be used for real-time visualization. We again obtain $\frac{\sqrt{5}}{2}$ as an upper bound for the wobbling distance. Needless to say, this engineering machinery was the original motivation for the development of the approach to wobbling bijections using shearings.

    Figure 3 shows the wobbling φ (4.12) for the angle $10°$. Figure 4 shows the upper bound $\frac{1}{2}\sqrt{3 + 2(\frac{1-t^2}{1+t^2}) + t^2}$ for the wobbling distances of rotated lattices. Moreover, we evaluated the wobbling bijection (4.12) for unit lattices of size 500×500 and 1000×1000 centred at the origin. The vertical bars indicate the achieved wobbling distances and the reasonably good empirical convergence to (4.9) with the usual number-theoretical background noise.

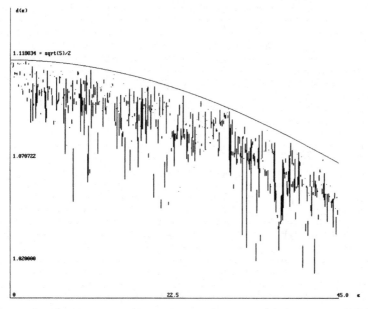

Figure 4

5. Applications of Sturmian sequences for wobbling bijections and billiards

5.1. Recursion formulas for some Sturmian sequences

We consider the lattice $\alpha \mathbb{Z} \times \frac{1}{\alpha}\mathbb{Z}$ and give a number-theoretical construction for a wobbling bijection to a standard lattice, following the ideas of E. Koppenrade (see [7]). This is motivated by the fact that the continued fractions of α and $1/\alpha$ are essentially the same, which suggests that there should be some deeper insight than just the results indicated in the previous sections. For related results see [2].

In billiards theory we encounter so-called *Sturmian sequences*. Playing billiards on a square table, one drives the ball from the lower left corner in the direction $\alpha = tan$ (*driving angle*). The ball will successively hit the 'horizontal' and 'vertical' edges. The sequence $S(\alpha)$ of letters h, v indicating the succession of the hits is the so-called Sturmian sequence of the billiard α.

Notation. Let A^* be the free monoid over a finite alphabet A with concatenation $*$ of the words; $\emptyset$ denotes the empty word and a^n is the n-fold concatenation of $a \in A^*$, where $a^0 = \emptyset$.

Definition 5.1. *Let $1/\alpha = [a_0, a_1, a_2, \ldots]$ be the continued fraction expansion of $1/\alpha$ with $a_0 \in \mathbb{N}_0$, $a_i \in \mathbb{N}$ for $i \geqslant 1$ and r_i a strictly decreasing sequence with $r_i < a_i$ as defined in Section 2. Let*

$$b_0 = a_0, \qquad b_1 = b_0^{a_1-1} * (a_0 + 1), \tag{5.1}$$

$$b_k = b_{k-1}^{a_k-1} * b_{k-2} * b_{k-1}, \qquad and \tag{5.2}$$

$$B(\alpha) = b_0 * b_1 * b_2 * b_3 * \cdots. \tag{5.3}$$

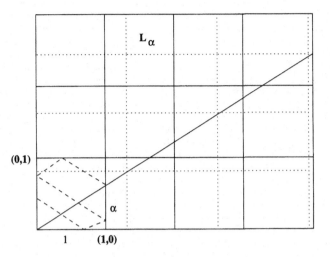

$$Z^2 = L_1$$

Figure 5

Observe that this yields a method for writing words rather efficiently, in fact exponentially fast. Computer linguists have been enthusiastic about the possibility of generating meaningful text exponentially fast. The classical Sturmian sequence for a billiard α (the tangent of the driving angle) $S(\alpha)$ is obtained from $B(\alpha)$ as

$$S(\alpha) = v^{B_0(\alpha)} * h * v^{B_1(\alpha)} * h * v^{B_2(\alpha)} * h * v^{B_3(\alpha)} \cdots, \tag{5.4}$$

where $B_i(\alpha)$ is the ith letter in $B(\alpha)$. Obviously there is a forgetful functor between $B(\alpha)$ and $S(\alpha)$.

A standard technique of unfolding billiards in the plane is shown in Figure 5.

In this way we see that the Sturmian sequence indicates how a straight line with slope α traverses the unit lattice. This, and its relation to the unit lattice as below, is relevant for the rectangular lattice $\alpha\mathbb{Z} \times \frac{1}{\alpha}\mathbb{Z}$, where the straight line is generated by one of the diagonals of the basic rectangle.

Definition 5.2. *Let $\alpha \in \mathbb{R}$ and $C(\alpha)$ be the sequence $C_k(\alpha) = |\alpha\mathbb{N} \cap (k, k+1]|$.*

Let W, W' be two finite segments of $C(\alpha)$ of the same length. Then we have the classical Sturmian property

$$\left| \sum_{x_i \in W} x_i - \sum_{x'_i \in W'} x'_i \right| \leqslant 1. \tag{5.5}$$

Theorem 5.1. $C(\alpha) = B(\alpha)$.

Theorem 5.1 gives the natural interpretation of the 'Sturmian sequence'. In the following, 'Sturmian sequence' refers to any of the three sequences $B(\alpha)$, $C(\alpha)$ or $S(\alpha)$.

The proof of Theorem 5.1 will be by induction using the definitions. We shall make use of shifted sequences.

Definition 5.3. *For $x \in \mathbb{R}$, $k \in \mathbb{N}$, let*

$$b(x,k) = |x + \alpha\mathbb{N} \cap (k, k+1]|. \tag{5.6}$$

Thus $b(x,k)$ tells us how many times the arithmetic progression $x + \alpha$, $x + 2\alpha, \ldots$ hits the interval $(k, k+1]$. Obviously $b(0,k) = C_k(\alpha)$ for every k, which will be useful for the proof of Theorem 5.1.

The following innocent-looking shifting lemma will be used frequently.

Lemma 5.1 (Shifting Lemma). *Let $x, z \in \mathbb{R}$ and $i, l \in \mathbb{N}_0$, such that $z \leqslant l$ and $z - x + (i - l) \in \alpha\mathbb{N}_0$. Then one has $b(z, l) = b(x, i)$.*

If one wants to evaluate b at some place, one can do so at a more convenient place provided that the hypotheses of the Shifting Lemma are observed.

Proof of the Shifting Lemma. Let x, z, i, l satisfy the hypotheses of the Shifting Lemma and let $(z - x) + (i - l) = p\alpha$ with $p \in \mathbb{N}_0$. Then one has

$$x + p\alpha = z - l + i \leqslant i. \tag{5.7}$$

Using Definition 5.3 and (5.7), one obtains $b(z, l) = |z + \alpha\mathbb{N} \cap (l, l+1]| = |z + p\alpha - z + x + \alpha\mathbb{N} \cap (i, i+1]| = |x + p\alpha + \alpha\mathbb{N} \cap (i, i+1]| = |x + \alpha\mathbb{N} \cap (i, i+1]| = b(x, i)$. ∎

Hereafter we shall have to verify the hypotheses of the Shifting Lemma quite often and it will be the technical routine part of the proofs. The hypothesis $z \leqslant l$ is certainly satisfied if $z \leqslant 0$, a fact we shall use without further mention.

Notation. Let

$$\beta(k) = \sum_{j=0}^{k-1} (-1)^{j+1} r_j \quad \text{for } k \in \mathbb{N}_0. \tag{5.8}$$

We need the following definition.

Definition 5.4. *For $k \in \mathbb{N}_0$ let*

$$
\begin{aligned}
u_k &:= \beta(k) - (-1)^k r_{k+1} - (a_{k+1} - 1)(-1)^k r_k \\
\text{and} \quad v_k &:= \beta(k) + (-1)^k r_k; \\
\text{let} \quad B_k &:= (\min\{u_k, v_k\}, \max\{u_k, v_k\}).
\end{aligned}
\tag{5.9}
\tag{5.10}
$$

Obviously $\beta(k) \in B_k$ and we shall call $\beta(k)$ the *base-point* of B_k. Since the sequence $r_0, r_1 \ldots$ is strictly decreasing, all base-points satisfy $\beta(k) \leqslant 0$, and for $k > 0$ the inequality is strict. For the same reason $\sup B_k \leqslant 0$ if $k > 0$ and $\sup B_0 = r_0$.

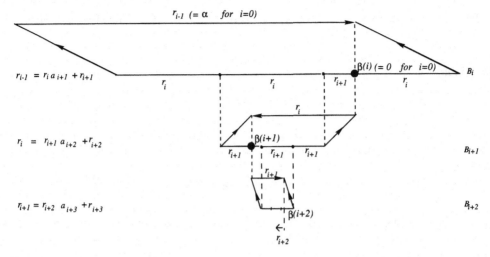

Figure 6

Notation. Let b_k be according to (5.1), (5.2). By $|b_k|$ we denote the length of the word b_k, by $b_{k,i}$ the ith entry, and by $\sum(b_k) = \sum_{i<|b_k|} b_{k,i}$ the sum over all entries in b_k.

Theorem 5.1 follows from the following lemmas.

Lemma 5.2. Let $k \in \mathbb{N}_0$. Then $\alpha \sum(b_k) = |b_k| + (-1)^{k+1} r_k$.

Lemma 5.3. Let $x \in B_k$ and $i < |b_k|$. Then $b(x,i) = b_{k,i}$.

Observe that Lemma 5.2 gives an indication of the quality of approximation of α in terms of the length of the subwords b_k of $B(\alpha)$. We have $\lim_{k\to\infty} \frac{|b_k|}{\sum(b_k)} = \alpha$, that is, the length of the kth word, divided by the sum of entries tends to α. This approximation is measured by the error term of the continued fraction r_{k+1} again divided by the sum of entries $\sum(b_k)$. This obviously relates to classical approximation via continued fractions.

Proof of Theorem 5.1. We shall evaluate the ith letter of the subword b_k of $B(\alpha)$. Using Lemma 5.3 and the Shifting Lemma, obtain

$$B_{i+\sum_{j=0}^{k-1}|b_j|}(\alpha) = b_{k,i} = b(\beta(k),i) = b\left(0,\left(i+\sum_{j=0}^{k-1}|b_j|\right)\right) = C_{i+\sum_{j=0}^{k-1}|b_j|}(\alpha). \qquad (5.11)$$

This shows that $B(\alpha) = C(\alpha)$, where the last equality follows from the remark following Definition 5.3.

We still have to verify the hypotheses of the Shifting Lemma. The first one is given by $\beta(k) \leqslant 0$. For the other, note that, by definition of $\beta(k)$ and Lemma 5.2, one has

$$(\beta(k) - 0) + \left(i + \sum_{j=0}^{k-1} |b_j| - i \right) = \beta(k) + \sum_{j=0}^{k-1} |b_j|$$

$$= \sum_{j=0}^{k-1} (-1)^{j+1} r_j + \sum_{j=0}^{k-1} |b_j|$$

$$= \sum_{j=0}^{k-1} \left((-1)^{j+1} r_j + |b_j| \right)$$

$$= \sum_{j=0}^{k-1} \left(\alpha \sum (b_j) \right) \in \alpha \mathbb{N}_0. \tag{5.12}$$

$\square$

Proof of Lemma 5.2. Proceed by induction on k. For $k = 0$ obtain:

$$\alpha \sum (b_0) = \alpha a_0 = 1 - r_0 = |b_0| + (-1)^{0+1} r_0. \tag{5.13}$$

For $k = 1$ we use the continued fraction expansion and definition (5.1):

$$
\begin{aligned}
|b_1| + r_1 &= (a_1 - 1) + 1 + r_1 \\
&= a_1 + r_1 \\
&= a_1 + \alpha - r_0 a_1 \\
&= a_1(1 - r_0) + \alpha \\
&= a_1(a_0 \alpha) + \alpha \\
&= \alpha(a_1 a_0 - a_0 + a_0 + 1) \\
&= \alpha((a_1 - 1)a_0 + (a_0 + 1)) \\
&= \alpha(a_1 - 1) \sum (b_0) + (a_0 + 1) \\
&= \alpha \sum (b_1).
\end{aligned}
$$

For the induction start with Definition 5.3: after some elementary transformations the induction hypothesis may be applied. By another elementary transformation and the definition of b_k using continued fractions, the lemma follows:

$$
\begin{aligned}
\alpha \sum (b_k) &= \alpha \left(a_k \sum (b_{k-1}) + \sum (b_{k-2}) \right) \\
&= \alpha a_k \sum (b_{k-1}) + \alpha \sum (b_{k-2}) \\
&= a_k(|b_{k-1}| + (-1)^k r_{k-1}) + (|b_{k-2}| + (-1)^{k-1} r_{k-2}) \\
&= a_k |b_{k-1}| + |b_{k-2}| + (-1)^{k+1}(r_{k-2} - a_k r_{k-1}) \\
&= |b_k| + (-1)^{k+1} r_k. \tag{5.14}
\end{aligned}
$$

$\square$

Proof of Lemma 5.3. Proceed by induction on k. For $k = 0$ we have $b_0 = a_0$. Thus we show that $b(x, 0) = a_0$ for $x \in B_0$. By (5.2) and (5.6) we show $|x + \alpha \mathbb{N} \cap (0, 1]| = a_0$ for $x \in B_0$. In order to do so let $x \in B_0$. By (5.10) for $k = 0$, one has

$$r_0 - \alpha < x < r_0; \tag{5.15}$$

thus $x + \alpha n > r_0 - \alpha + \alpha = r_0 > 0$ for all $n \in \mathbb{N}$. Consider the smallest n_0 such that $x + \alpha n_0 > 1$, that is, $x + \alpha n_0 \notin (0, 1)$, but $x + \alpha(n_0 - 1) \in (0, 1]$ provided $n_0 - 1 \in \mathbb{N}$. By (5.15) obtain $x + \alpha a_0 < r_0 + \alpha a_0 = 1$, and $x + \alpha(a_0 + 1) > r_0 - \alpha + \alpha(a_0 + 1) = r_0 + \alpha a_0 = 1$. So we have $n_0 = a_0 + 1$, which shows that $x + \alpha \mathbb{N} \cap (0, 1] = \{x + \alpha, \dots, x + a_0\alpha\}$; thus $b(x, 0) = a_0$.

For $k = 1$ let $x \in B_1$, and remember that (5.1) states that $b_1 = a_0^{a_1 - 1} * (a_0 + 1)$. Thus we consider two cases.

Case I. Let $0 \leq i < a_i - 1$ and consider the ith letter of b_1. We want to show that $b(x, i) = b_{1,i} = a_0$ for these is. As $x \in B_1$, (5.10) yields $x - ir_0 \in B_0$. Therefore we can apply Lemma 5.3 with $k = 0$ and obtain by the Shifting Lemma (Lemma 5.1) $b(x, i) = b(x - ir_0, 0) = a_0$ as desired. It remains to check the hypotheses of the Shifting Lemma.

- For the hypothesis '$z \leq l$', observe that $\sup B_1 \leq 0$, as noted in the definition of the B_is.
- For the other hypothesis one has $x - ir_0 - x + i = i - ir_0 = i(1 - r_0) = i\alpha a_0$.

Case II. For $i = a_1 - 1$, we observe that $x - (a_1 - 1)r_0 + \alpha \in B_0$; shifting the arguments by $a_1 - 1$ and applying continued fractions, we obtain

$$
\begin{aligned}
a_0 + 1 &= b(x - (a_1 - 1)r_0 + \alpha, 0) + 1 \\
&= b(x + (a_1 - 1)(1 - r_0) + \alpha, a_1 - 1) + 1 \\
&= b(x + a_0(a_1 - 1)\alpha + \alpha, a_1 - 1) + 1 \\
&= b(x + a_0(a_1 - 1)\alpha, a_1 - 1) \\
&= b(x, a_1 - 1). \tag{5.16}
\end{aligned}
$$

So we get $a_0 + 1 = b(x + a_0(a_1 - 1)\alpha, a_1 - 1) = b(x, a_1 - 1)$ by the Shifting Lemma, since $a_1 \geq 1$ and

$$x + a_0(a_1 - 1)\alpha - (a_1 - 1) < -r_1 + (a_1 - 1)(a_0\alpha - 1) - r_1 - r_0(a_1 - 1) \leq -r_1 \leq 0.$$

Now assume that Lemma 5.3 holds for all integers less than $k + 2$ for some $k \geq 0$. Fix $x \in B_{k+2}$ and recall that $b_{k+2} = b_{k+1}^{a_{k+2}} * b_k * b_{k+1}$; thus $|b_{k+2}| = (a_{k+2} - 1)|b_{k+1}| + |b_k| + |b_{k+1}|$. For $i < |b_{k+2}|$ one of the following cases applies:

(i)	$0 \leq i$	$=$	$|b_{k+1}|j + l$,	where	$0 \leq j < a_{k+2} - 1$, $\ 0 \leq l < |b_{k+1}|$,
(ii)	i	$=$	$|b_{k+1}|(a_{k+2} - 1) + l$,	where	$0 \leq l < |b_k|$,
(iii)	i	$=$	$|b_{k+1}|(a_{k+2} - 1) + |b_k| + l$,	where	$0 \leq l < |b_{k+1}|$.

$$\tag{5.17}$$

Table 1

	α			$1/\alpha$	
b_0	$=$	a_0	b'_0	$=$	0
b_1	$=$	$a_0^{a_1-1} * (a_0 + 1)$	b'_1	$=$	$0^{a_0-1} * 1$
$\vdots$			$\vdots$		
b_k	$=$	$b_{k-1}^{a_k-1} * b_{k-2} * b_{k-1}$	b'_k	$=$	$b_{k-1}'^{a_k-1-1} * b'_{k-2} * b'_{k-1}$
$B(\alpha)$	$=$	$b_0 * b_1 * \cdots$	$B(1/\alpha)$	$=$	$b'_0 * b'_1 * \cdots$

By construction of B_k, B_{k+1}, B_{k+2}, observe that in each case

(i) $0 > x + (-1)^{k+2} r_{k+1} j$ $= x_1 \in B_{k+1}$,

(ii) $0 > x + (-1)^{k+2} r_{k+1}(a_{k+2} - 1)$ $= x_2 \in B_k$,

(iii) $0 > x + (-1)^{k+2} r_{k+1}(a_{k+2} - 1) + (-1)^{k+1} r_k = x_3 \in B_{k+1}$.

Using the induction hypothesis we show that

$$
\begin{array}{l}
\text{(i)} \\
\text{(ii)} \\
\text{(iii)}
\end{array}
\qquad
b_{k+2,i} = \left\{
\begin{array}{lll}
b_{k+1,l} & = & b(x_1, l) \\
b_{k,l} & = & b(x_2, l) \\
b_{k+1,l} & = & b(x_3, l)
\end{array}
\right\} = b(x, i),
\qquad (5.18)
$$

for each case, respectively. As $x_m < 0 \leqslant l$ for $m \in \{1, 2, 3\}$, the first hypothesis of the Shifting Lemma is verified. The proof is complete provided that we can verify the other hypothesis of the Shifting Lemma:

$$(x_m - x) + (i - l) \in \alpha \mathbb{N}_0 \quad \text{for} \quad m = 1, 2, 3. \qquad (5.19)$$

We prototypically consider case (iii) and leave the rest to the reader.

$$
\begin{aligned}
(x_3 - x) + (i - l) &= (-1)^{k+2} r_{k+1}(a_{k+2} - 1) + (-1)^{k+1} r_k + |b_{k+1}|(a_{k+1} - 1) + |b_k| \\
&= (a_{k+2} - 1)((-1)^{k+2} r_{k+1} + |b_{k+1}|) + (-1)^{k+1} r_k + |b_k| \\
&= \alpha(a_{k+2} - 1) \sum (b_{k+1}) + \alpha \sum (b_k)
\end{aligned}
$$

by Lemma 5.2. But the last term is in $\alpha \mathbb{N}_0$ as desired. This completes the proof of Lemma 5.3 and thus the proof of Theorem 5.1.   □

5.2. Wobbling bijections for rectangular lattices

In this section we shall give an 'explicit' wobbling bijection between the rectangular lattice $R = \alpha \mathbb{N} \times 1/\alpha \mathbb{N}$ and $\mathbb{N}^2$ using the Sturmian sequence.

For $\alpha < 1$ let $\alpha = [0, a_0, a_1, a_2, \ldots]$ be the continued fraction expansion. The Sturmian sequences $B(\alpha)$ and $B(1/\alpha)$ are shown in Table 1. Note that $B(\alpha)$ is over the alphabet $\{a_0, a_0 + 1\}$ and $B(1/\alpha)$ is over the alphabet $\{0, 1\}$. One easily obtains $B(1/\alpha)$ from $B(\alpha)$

by means of the following *rewriting system*:

$$f(a_0) = 0^{a_0-1} * 1,$$
$$f(a_0 + 1) = 0^{a_0} * 1. \tag{5.20}$$

Then one obtains

$$B(1/\alpha) = 0 * f(B(\alpha)).$$

Define the following infinite $0, \pm 1$ matrix $N = (n_{i,j})_{i,j}$, where $n_{i,j} = B_i(\alpha) - B_j(\alpha)$ and $B_i(\alpha)$ denotes the ith term in the sequence $B(\alpha)$.

Lemma 5.4. *Along all diagonals of N, the elements $+1$ and -1 occur alternately.*

Proof. Assume that $n_{i,j}$ and $n_{i+k,j+k}$ are two consecutive nonzero entries (the case $n_{i+k,j-k}$ can be handled similarly). Say $n_{i,j} = +1$. Then, by definition,

$$B_i(\alpha) \neq B_j(\alpha)$$
$$B_{i+1}(\alpha) = B_{j+1}(\alpha)$$
$$\vdots \qquad \vdots$$
$$B_{i+k-1}(\alpha) = B_{j+k-1}(\alpha)$$
$$B_{i+k}(\alpha) = ? \tag{5.21}$$

The words $W = (B_i(\alpha), \ldots, B_{i+k}(\alpha))$ and $W' = (B_j(\alpha), \ldots, B_{j+k}(\alpha))$ are segments of $B(\alpha)$ of the same length k. By the Sturmian property $|\sum(W) - \sum(W')| \leq 1$. Then $B_{i+k}(\alpha) = B_{j+k}(\alpha) - 1$, for otherwise $|\sum(W) - \sum(W')| = 2$ and $n_{i+k,j+k} = -1$, as desired. $\square$

We need the tiling

$$\mathbb{R}^+ \times \mathbb{R}^+ = \left(\bigcup_{j=0}^{\infty} [0,1) \times [j, j+1) \right) \dot{\cup} \left(\bigcup_{i,j \in \mathbb{N}} I_i \times [j, j+1) \right), \tag{5.22}$$

where (I_i) is a sequence of consecutive intervals of length $l(I_i) = B_i(\alpha)$.

Lemma 5.5. *For every i, j, the tile $I_i \times [j, j+1)$ intersects the lattice R in $B_j(\alpha)$ many points.*

Proof. I_i has length $B_i(\alpha)$. By means of the rewriting definition (5.20) of $B(1/\alpha)$, the segment I_i of $B(1/\alpha)$ has exactly one entry of value one. Moreover, by Theorem 5.1 the Sturmian sequence $B_j(\alpha)$ counts $|\alpha \mathbb{N} \cap (j, j+1)|$. $\square$

Now we define a wobbling bijection between R and $\mathbb{N} \times \mathbb{N}$, as follows. Consider a tile $T = I_i \times [j, j+1)$ with the property $n_{i,j} = +1$ (say). By definition,

$$1 + |T \cap \mathbb{N}^2| = |T \cap R|; \tag{5.23}$$

thus there is one vertex too many to allow a local bijection between parts of R and $\mathbb{N}^2$

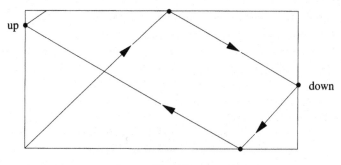

Figure 7

within T. Now successively shift one surplus point along the diagonal into the next tile. After suitably many iterations, according to Lemma 5.4 this procedure ends in a tile T' with $n_{i,j} = -1$ balancing the process. Finally this local procedure gives a global bijection by patching together the local mappings.

This is a wobbling mapping, as one maps from any tile either to itself or to an adjoining diagonal neighbouring tile, and tiles have size at most $(a_0 + 1) \times 1$. For the particular choice of $R = \frac{1}{\alpha}\mathbb{N} \times \alpha\mathbb{N}$ it should be technically possible to calculate reasonable wobbling distances [10].

5.3. Discrepancies of billiards

Here we explore the Sturmian sequence for a very special billiard $\alpha = \sqrt{2}$. This can surely be generalized, but this simple case is illustrative. Consider the billiard on a square table starting the ball in the lower left corner with angle of slope α. Then the ball will alternately hit the left and right vertical edges. In order to obtain a more sophisticated problem we consider the sequence of 'upwards' and 'downwards' hits of vertical edges and ask for the discrepancy as shown in Figure 7.

Let $\bigcup_N(\alpha)$ be the number of upward hits among the first N vertical hits, and similarly $\bigcap_N(\alpha)$ for the downward hits. The discrepancy will be

$$D_N(\alpha) = \bigcup_N(\alpha) - \frac{N}{2} = \frac{\bigcup_N(\alpha) - \bigcap_N(\alpha)}{2}. \tag{5.24}$$

We only want to consider this discrepancy for very special Ns. Consider the Sturmian sequence and let N be of the form

$$N = \sum (b_0 * \cdots * b_k) = \sum_{i < \text{length}(b_0 * \cdots * b_k)} B_i(\alpha) \tag{5.25}$$

for some k. Obviously, by (5.3) $k \sim \log N$. Then, from (5.4),

$$\bigcup_N(\alpha) = B_0(\alpha) + B_2(\alpha) + B_4(\alpha) + \cdots,$$

$$\bigcap_N(\alpha) = B_1(\alpha) + B_3(\alpha) + \cdots, \tag{5.26}$$

where sums extend over all even (odd) $i < \text{length}(b_0 * \cdots * b_k)$. Therefore

$$\bigcup_N(\alpha) - \bigcap_N(\alpha) = B_0(\alpha) - B_1(\alpha) + B_2(\alpha) - B_3(\alpha) \ldots . \tag{5.27}$$

In order to simplify the further calculations we take $\alpha = \sqrt{2} = [1, \bar{2}, \ldots]$. One has $b_0 = 0$, $b_1 = 1$, $b_2 = 101$ and $b_k = b_{k-1} * b_{k-2} * b_{k-1}$.

Notation. Let $x = (x_0 \ldots x_l)$; then $\overline{\sum}(x) = \sum (-1)^j x_j$.

Lemma 5.6. *For* $\alpha = \sqrt{2}$,

$$U_N(\alpha) - \bigcap_N(\alpha) = \sum_{j \leq k} \left((-1)^j \overline{\sum}(b_j) \right). \tag{5.28}$$

Proof. Proceed by induction on k. For $k = 0, 1$ the statement obviously holds. Assume that it holds for some $k \geq 1$. Then

$$
\begin{aligned}
U_{N(k+1)}(\sqrt{2}) - \bigcap_{N(k+1)}(\sqrt{2}) &= U_{N(k)}(\sqrt{2}) - \bigcap_{N(k)}(\sqrt{2}) + (-1)^{k+1} \overline{\sum}(b_{k+1}) \\
&= \sum_{j \leq k} \left((-1)^j \overline{\sum}(b_j) \right) + (-1)^{k+1} \overline{\sum}(b_{k+1}) \\
&= \sum_{j \leq k+1} \left((-1)^j \overline{\sum}(b_j) \right). \tag{5.29}
\end{aligned}
$$

$\square$

Lemma 5.7. *For* $\alpha = \sqrt{2}$ *one has, for all* $j \in \mathbb{N}_0$,

$$\overline{\sum}(b_j) = j. \tag{5.30}$$

Proof. Proceed by induction on j. The cases $j = 0, 1$ are obvious. For $\alpha = \sqrt{2}$ the recursion (5.2) specializes to $b_j = b_{j-1} * b_{j-2} * b_{j-1}$, using the fact that $|b_j| = 1 \pmod 2$ for all j. The induction step shows

$$\overline{\sum}(b_j) = 2 \overline{\sum}(b_{j-1}) - \overline{\sum}(b_{j-2}) = j. \tag{5.31}$$

Combining these facts, obtain

$$U_N(\sqrt{2}) - \bigcap_N(\sqrt{2}) = \sum_{j \leq k} \left((-1)^j \overline{\sum}(b_j) \right) = \sum_{j \leq k} (-1)^j \cdot j = \begin{cases} +k & \text{for } k \text{ even,} \\ -(k+1) & \text{for } k \text{ odd.} \end{cases} \tag{5.32}$$

$\square$

Corollary 5.1. *For the* $\sqrt{2}$-*billiard the discrepancy* $D_N(\sqrt{2})$ *is at least* $\log N$ *for infinitely many* N. $\square$

This is a very special case of a deep theorem by Roth, stating that discrepancies are not too small. Here we considered the simple billiard model in order to have an elementary, accessible example.

6. Paradoxical situations

A classical result by Tarski [1] states that the unit ball B in $\mathbb{R}^3$ may be decomposed into two parts B_1, B_2 such that B, B_1, B_2 are pairwise congruent. Paradoxical decompositions were defined for metric spaces with wobbling equivalences [5].

Definition 6.1. *Let M be a metric space. M is* paradoxical *if there is a decomposition into two parts M_1, M_2 such that M, M_1, M_2 are pairwise equivalent.*

For discrete countable metric spaces, one can define exponential richness via the concept of the 'doubling radius' r, as follows.

Definition 6.2. *Let M be a countable discrete metric space. M is* at least exponentially rich *if there exists $r \in \mathbb{R}^+$ such that, for every finite subset X of M, the r-neighbourhood of X satisfies*

$$|N_r(X)| \geqslant 2|X|. \tag{6.1}$$

Theorem 6.1 ([5]). *Let M be a discrete countable metric space. Then M is paradoxical if M is at least exponentially rich.*

Let us report that Elek [6] used this theorem in order to characterize the finitely generated amenable groups.

Theorem 6.1 is as effective as Hall and Rado's Matching Theorem, which is used for the proof of the Hall–Rado theorem. A recursive version does not hold [9], but paradoxicality can be tested by countably many tests on finite sets.

Let X be a recursive subset of $\mathbb{Q}^2$, and d be a positive real. If X has a d-wobbling to $\mathbb{Z}^2$, is there a recursive one? For trivial reasons, the answer is 'yes' if d is less than $\frac{1}{2}$. Let us show that the answer in general is 'no' if $d = \frac{1}{2}$.

Theorem 6.2. *There is a recursive set $X \subseteq \mathbb{Q}^2$ having a $\frac{1}{2}$-wobbling to $\mathbb{Z}^2$, but no recursive one.*

Proof. The proof will be a classical diagonalization argument. Every index e_i is associated with the vertical line L_i through $(i, 0)$. The aim is to construct X in such a way that φ_{e_i}, the algorithm having index e_i, working wrongly on L_i, cannot be a $\frac{1}{2}$-wobbling of X to $\mathbb{Z}^2$. The set X is defined inductively by stages:

$$X_0 \subseteq X_1 \subseteq \cdots \subseteq X_n \subseteq \cdots \qquad \text{such that } \bigcup_n X_n = X. \tag{6.2}$$

For book-keeping purposes we use marker and lists Λ_n of 'inactive' indices. Let $e_1, e_2, e_3, \ldots$ be a recursive enumeration of indices of all partial recursive functions on $\mathbb{Q}^2$ to $\mathbb{Z}^2$.

stage 0:
$$X_0 := \{(a,b) \mid a,b \in \mathbb{Z} \wedge a \leqslant 0\}$$
$$\cup \quad \{(i,\tfrac{1}{2}) \mid i \in \mathbb{N} \setminus \{0\}\}$$
$$\cup \quad \{(i,-\tfrac{1}{2}) \mid i \in \mathbb{N} \setminus \{0\}\},$$
$$\Lambda_0 := \emptyset.$$

X_0, Λ_0 are recursive sets.

stage $n+1$:   We consider $\{e_1,\ldots,e_{n+1}\} \setminus \Lambda_n =: A_{n+1}$,

the set of indices active in stage $n+1$.

Let e_i be an element of A_{n+1}. We compute $n+1$ steps of φ_{e_i} for argument $(i,\tfrac{1}{2})$.

case $n+1$ (1): $\varphi_{e_i}^{n+1}$ gives output $(i,1)$.   We put a marker on $(i,-(n+1)+\tfrac{1}{2})$.

case $n+1$ (2): $\varphi_{e_i}^{n+1}$ gives output $(i,0)$.   We put a marker on $(i,n+1+\tfrac{1}{2})$.

case $n+1$ (3): $\varphi_{e_i}^{n+1}$ gives no output, or   We place no marker.
does something else.

We define

$$X_{n+1} := X_n \cup \{(a,b+\tfrac{1}{2}) \mid a \in \mathbb{N} \text{ and } 1 \leqslant a \leqslant n+1 \text{ and } b \in \mathbb{Z}$$
$$\text{and } -(n+1) \leqslant b \leqslant n+1 \text{ and } (a,b+\tfrac{1}{2}) \text{ unmarked}\}. \tag{6.3}$$

Obviously, X_{n+1} is recursive. $\Lambda_{n+1} := \Lambda_n \cup \{e_i \in A_{n+1} \mid \varphi_{e_i}^{n+1} \text{ gives an output}\}$. Clearly $X = \bigcup_n X_n$ is recursive.

Let us show that there is a $\tfrac{1}{2}$-wobbling φ of X to $\mathbb{Z}^2$. We consider the vertical lines L_{-i} for $i \in \mathbb{N}$. On these lines φ is the identity. We consider the vertical lines L_i, $1 \leqslant i$. On these lines φ works as follows. If there is no marker on line L_i we define $\varphi(i,z+\tfrac{1}{2}) = (i,z)$, $z \in \mathbb{Z}$. If a marker is on $(i,-(n+1)+\tfrac{1}{2})$, we define

$$\varphi(i,z+\tfrac{1}{2}) = (i,z), \quad z \geqslant -n \qquad \text{and} \qquad \varphi(i,z+\tfrac{1}{2}) = (i,z+1), \quad z < -n. \tag{6.4}$$

If a marker is on $(i,n+1+\tfrac{1}{2})$, we define

$$\varphi(i,z+\tfrac{1}{2}) = (i,z+1), \quad z \leqslant n \qquad \text{and} \qquad \varphi(i,z+\tfrac{1}{2}) = (i,z), \quad z > n. \tag{6.5}$$

Obviously $\varphi(X) = \mathbb{Z}^2$, and φ is injective. So φ is a $\tfrac{1}{2}$-wobbling of X onto $\mathbb{Z}^2$.

Now we prove that no φ_{e_i} is a $\tfrac{1}{2}$-wobbling of X to $\mathbb{Z}^2$. Suppose that φ_{e_i} is a $\tfrac{1}{2}$-wobbling. Then φ_{e_i} is eventually defined on $(i,\tfrac{1}{2})$, say in step $n+1$. Any $\tfrac{1}{2}$-wobbling of X to $\mathbb{Z}^2$ moves only at vertical lines. If $\varphi_{e_i}^{n+1}(i,\tfrac{1}{2}) = (i,1)$, a marker is on $(i,-(n+1)+\tfrac{1}{2})$. Hence $(i,-(n+1)+\tfrac{1}{2})$ is not in X_{n+1} and *a fortiori* not in X. To close the gap at $(i,-(n+1)+\tfrac{1}{2})$, any $\tfrac{1}{2}$-wobbling is forced to work like φ above. So φ_{e_i} moves $(i,\tfrac{1}{2})$ in the wrong direction. If $\varphi_{e_i}^{n+1}(i,\tfrac{1}{2}) = (i,0)$ the same argument applies. Hence there is no recursive $\tfrac{1}{2}$-wobbling of X to $\mathbb{Z}^2$.   □

Remark. The above argument is prototypical and may be adapted to show that for every $d \geqslant 1/2$ the analogue of Theorem 6.2 holds. A general method to construct counterexamples such as the one above is given in [3].

References

[1] Banach, S. and Tarski, A. (1924) Sur la decomposition des ensembles de points en parties respectivement congruents. *Fund. Math.* **6** 244–277.

[2] Brown, T. C. (1993) Description of the characteristic sequence of an irrational. *Canad. Math. Bull.* **36** 15–21.

[3] Carstens, H.-G. and Päppinghaus, P. (1984) Abstract construction of counterexamples in recursive graph theory. In *Computation and Proof Theory* (M. M. Richter, E. Börger, W. Oberschelp, B. Schinzel, and W. Thomas, eds), Vol. 1104 of *Lecture Notes in Mathematics*, Springer.

[4] Ceccherini-Silberstein, T., Grigorchuk, R. and de la Harpe, P. (1997) Amenability and paradoxes for pseudogroups and for discrete metric spaces. Manuscript, personal communication.

[5] Deuber, W. A., Simonovits, M. and Sós, V. T. (1995) A note on paradoxical metric spaces. *Studia Scientiarum Mathematicarum Hungarica* **30** 17–23.

[6] Elek, G. (1997) The K-theory of Gromov's translation algebras and the amenability of discrete groups. *Proc. Amer. Math. Soc.* **125** 2551–2553.

[7] Koppenrade, E. (1996) Beiträge zum Studium effektiver distanzbeschränkter Abbildungen zwischen Gittern mit besonderer Berücksichtigung rotierter Gitter unter Verwendung der Kettenbruchdarstellung. Dissertation, Universität Bielefeld.

[8] Laczkovich, M. (1990) Equidecomposability and discrepancy: a solution of Tarski's circle-squaring problem. *J. Reine Angew. Math.* **404** 77–117.

[9] Manaster, A. and Rosenstein, J. (1972) Effective matchmaking (recursion theoretic aspects of a theorem of Philip Hall). *Proc. London Math. Soc.* **25** 615–654.

[10] Sudmeier, R. (1997) Beitrag zum Studium distanzbeschränkter Bijektionen zwischen Laczkovich-äquivalenten Mengen unter besonderer Berücksichtigung von Rechteckgittern. Diplomarbeit, Universität Bielefeld.

Combinatorics, Probability and Computing (1999) 8, 131–159.
© 1999 Cambridge University Press

On Random Intersection Graphs:
The Subgraph Problem

MICHAŁ KAROŃSKI[1]†, EDWARD R. SCHEINERMAN[2]‡

and KAREN B. SINGER-COHEN[3]‡

[1] Faculty of Mathematics and Computer Science,
Adam Mickiewicz University, Poznań, Poland
and
Department of Mathematics and Computer Science,
Emory University, Atlanta, GA 30322, USA
(e-mail: karonski@math.amu.edu.pl and michal@mathcs.emory.edu)

[2] Department of Mathematical Sciences, The Johns Hopkins University,
Baltimore, MD 21218–2689, USA
(e-mail: ers@cs.jhu.edu)

[3] Department of Mathematical Sciences, The Johns Hopkins University,
Baltimore, MD 21218–2689, USA
and
School of Mathematics, University of Minnesota,
Minneapolis, MN 55455, USA
(e-mail: singer@math.umn.edu)

A new model of random graphs – *random intersection graphs* – is introduced. In this model, vertices are assigned random subsets of a given set. Two vertices are adjacent provided their assigned sets intersect. We explore the evolution of random intersection graphs by studying thresholds for the appearance and disappearance of small induced subgraphs. An application to gate matrix circuit design is presented.

1. Introduction

1.1. The model

In most models of random graphs, the edges enjoy all the attention and the vertices are passive bystanders. In Erdős–Rényi random graph theory, we are given n vertices, and flip coins to see where the edges go – the appearance of one edge is independent of any other. Such a model is useful when the 'relations' between 'objects' are independent of one another.

† Research supported in part by the Komitet Badań Naukowych, grant 2 P03A 023 09.
‡ Research supported in part by the Office of Naval Research.

In this paper, we explore a model of random graphs in which the vertices are the focus. We independently assign to each vertex a random structure and then assess the adjacency of two vertices by comparing their assigned structures. To do this, we use the concept of an *intersection graph*.

Let G be a (finite, simple) graph. We say that G is an *intersection graph* if we can assign to each vertex $v \in V(G)$ a set S_v so that $vw \in E(G)$ (we write $v \sim w$) exactly when $S_v \cap S_w \neq \emptyset$. In this case, we say G is the intersection graph of the family of sets $\mathscr{S} = \{S_v : v \in V(G)\}$. It is easy to check that every graph is an intersection graph (see [15]).

If one restricts the choices for the sets S_v, various classes of graphs can be defined; the best known example is the class of *interval graphs*, in which the S_v must be real intervals. (See [17, 18] for a discussion of *random interval graphs*.)

We are now ready to define *random intersection graphs*. Let n, m be positive integers and let $p \in [0, 1]$. For every positive integer k with $1 \leqslant k \leqslant n$, let S_k be a random subset of $M = \{1, 2, \ldots, m\}$ formed by selecting each element of M independently with probability p. Thus the probability that we choose a particular set S for k is $p^s(1 - p)^{m-s}$, where $s = |S|$. Finally, let $G(n, m, p)$ be the intersection graph of the S_ks.

Thus $G(n, m, p)$ has n vertices $\{1, 2, \ldots, n\}$. We assign to each vertex k a random subset (as described above) $S_k \subseteq \{1, 2, \ldots, m\}$ and we have $i \sim j$ if and only if $S_i \cap S_j \neq \emptyset$.

Now, given two vertices u and v of $G(n, m, p)$ the probability that there is an edge connecting them is $\Pr\{u \sim v\} = 1 - (1 - p^2)^m$, since the probability that S_u and S_v are disjoint is simply $(1 - p^2)^m$. It follows that the expected number of edges in $G(n, m, p)$ is $\binom{n}{2}[1 - (1 - p^2)^m] \asymp n^2 mp^2$ (provided $mp^2 \to 0$ as $n \to \infty$). If we take $p = 1/(\omega_n n \sqrt{m})$, where ω_n denotes hereafter a function that goes to infinity with n, then the expected number of edges goes to 0 in the limit and with high probability[1] $G(n, m, p)$ is edgeless. Further, it follows from our results below that, when $p = \omega_n/(n\sqrt{m})$, then with high probability $G(n, m, p)$ has edges.

On the other hand, the expected number of *non-edges* in $G(n, m, p)$ is $\binom{n}{2}(1 - p^2)^m \asymp n^2 \exp\{-mp^2\}$. Thus, if we take $p = \sqrt{\frac{2\log n + \omega_n}{m}}$, then with high probability $G(n, m, p)$ is a complete graph. Further, when $p = \sqrt{(2\log n - \omega_n)/m}$, we show below that with high probability $G(n, m, p)$ is not complete.

Thus we may restrict our attention to values of p in the range between $1/(n\sqrt{m})$ and $\sqrt{2\log n/m}$. As p increases from the former to the latter, we witness the evolution of the structure of $G(n, m, p)$.

An alternative view of random intersection graph generation is given by its *representation matrix* $R(n, m, p)$. This matrix is an $n \times m$ matrix whose rows represent the vertices of $G(n, m, p)$ and whose columns represent the elements of the universal set $M = \{1, \ldots, m\}$. The entries in $R(n, m, p)$ are 0s and 1s; each entry is independently 1 with probability p (and 0 with probability $1 - p$). From the random representation matrix $R(n, m, p)$ we derive the graph $G(n, m, p)$ by deeming two vertices to be adjacent if and only if their

---

[1] As is customary in random graph theory, by *with high probability* we mean that the probability that $G = G(n, m, p)$ has the stated property tends to 1 as $n \to \infty$.

corresponding rows have a 1 in a common column (*i.e.*, their dot product is nonzero). Note that a given graph G may arise from many different representation matrices.

More formally, we let $\mathscr{R}(n, m, p)$ denote the sample space of all $n \times m$ 0,1-matrices with the probability of a particular matrix set to $p^a(1 - p)^b$, where a is the number of 1s and b is the number of 0s in the particular matrix. Now we let $\mathscr{G}(n, m, p)$ be the sample space of all graphs on n labelled vertices $\{1, 2, \ldots, n\}$. The probability of a particular graph G in $\mathscr{G}(n, m, p)$ is the sum of the probabilities of all matrices in $\mathscr{R}(n, m, p)$ that represent G.

Thus the random intersection graph $G(n, m, p)$ is an element of the sample space $\mathscr{G}(n, m, p)$.

We have seen how $G = G(n, m, p)$ arises from $R = R(n, m, p)$ by concentrating on the *rows* of R. A dual view of G is afforded by examining the *columns* of R. Consider a given column of R. The 1s in this column correspond to a collection of pairwise adjacent vertices in G, that is, a *clique*[2] of G. This said, we can think of $G(n, m, p)$ as being generated by the following random process. For $j = 1, \ldots, m$ do the following. Let C_j be a random subset of $V(G) = \{1, \ldots, n\}$ with each $i \in C_j$ independently with probability p. Having generated the sets $\{C_1, C_2, \ldots, C_m\}$, we declare vertices u and v to be adjacent exactly when they are together in a common C_j. In other words, the family $\mathscr{C} = \{C_1, C_2, \ldots, C_m\}$ is a *clique cover* of G, that is, a family of cliques of G with the property that every edge of G is induced in at least one of the C_js. The concept of clique covers is central to our discussion of subgraphs of $G(n, m, p)$.

The representation matrix $R(n, m, p)$ has yet another interpretation. In fact it can be viewed as the adjacency matrix of a random binomial bipartite graph $B(n, m, p)$ in which edges occur independently between vertices in the two parts $N = \{1, 2, \ldots, n\}$ and $M = \{1, 2, \ldots, m\}$ with probability p. A random intersection graph with the vertex set N is recovered from $B(n, m, p)$ as follows. We put an edge between two vertices u and v of $G(n, m, p)$ if and only if there is a vertex z in the M-part of $B(n, m, p)$ such that both $\{u, z\}$ and $\{v, z\}$ are edges of $B(n, m, p)$. Such an approach provides a useful relationship between the classical binomial model of a random graph with independent edges and the random intersection graph $G(n, m, p)$ where the edges are no longer independent.

We are interested in studying the properties of $G(n, m, p)$ for n large. We therefore have two 'parameters' that we can adjust: m and p. As we discuss below, when m is very small compared to n, the model is not particularly interesting, and when m is exceedingly large (compared to n) the behavior of $G(n, m, p)$ is essentially the same as for an Erdős–Rényi random graph. The 'right' balance is achieved when we take $m = \lfloor n^\alpha \rfloor$ where α is a positive constant, and this is the m we shall use. (From now on we drop the $\lfloor \cdot \rfloor$.)

1.2. Overview of results and applications

In Erdős–Rényi random graph theory, a basic question concerns the appearance of subgraphs during the evolution of a random graph. In particular, considering a fixed graph H, we ask the question: for which values of p is H with high probability an (induced) subgraph of the random graph? The answer depends on the *maximum average*

---

[2] A *clique* is a set of pairwise adjacent vertices. Cliques, for us, are not necessarily maximal cliques.

degree of H (the maximum of the average degree of all of H's subgraphs). For an overview, see [1, 5, 9].

One way to state the classical result is that a random graph $G(n, p)$ contains H with high probability exactly when the expected number of copies of H and all its subgraphs all go to infinity.

This paper studies the analogous problem for random intersection graphs. We show that for a fixed graph H there are two thresholds, τ_1 and τ_2. We show that $H \leqslant G(n, m, p)$ (we write $H \leqslant G$ to mean H is an induced subgraph of G) exactly when p is asymptotically between these thresholds. One of the curious special features of our model is that – in contradistinction to the Erdős–Rényi model – it is possible that the expected number of copies of H and all its subgraphs is very large, but the probability that $H \leqslant G(n, m, p)$ is very small.

We apply our results on subgraphs to answer the question: when is $G(n, m, p)$ with high probability an interval graph? In Section 3 we explain the importance of this question.

We believe that this 'vertex-biased' approach to random graphs can have important applications. Often the relationship between two 'objects' is not independent of the pair of objects. Objects that are 'closer' might be more likely to be related. For example, physical proximity is important in the spread of disease; the probability that a disease spreads from person A to person B is *not* independent of the two people chosen.

An application scenario well suited to our model of random graphs involves processors in a distributed setting. These processors 'compete' for shared resources (such as disks, printers, pages of memory, *etc.*) If each processor is oblivious to the actions of the others, then a reasonable protocol for the processor to follow is to try to secure its resources by making random selections. The graph $G(n, m, p)$ nicely models this situation. The n vertices are the processors, the m elements of the universal set are the resources, and processor i selects resource j with probability p. The edges of the resulting graph $G(n, m, p)$ represent the resulting pairwise conflicts that need to be resolved.

Another application of our model is to the *gate matrix layout* problem, which is discussed more fully in Section 3.

1.3. Probabilistic lemmas
Our proofs rely on the following probability results.

Lemma 1. *Let t be a fixed positive integer and let E denote an experiment with $t + 1$ mutually exclusive possible outcomes $\{0, \ldots, t\}$. Let p_j denote the probability that we observe outcome j. We perform n independent (with $n \to \infty$) trials of this experiment and let N_j denote the number of times we observe outcome j. Furthermore, suppose that for $1 \leqslant j \leqslant t$ we have $p_j \to 0$ as $n \to \infty$ (hence $p_0 \to 1$). Finally, let $a_1, a_2, \ldots, a_t$ be fixed, nonnegative integers. Then*

$$\frac{\Pr\{N_1 = a_1 \wedge N_2 = a_2 \wedge \cdots \wedge N_t = a_t\}}{\Pr\{N_1 = a_1\} \Pr\{N_2 = a_2\} \cdots \Pr\{N_t = a_t\}} \to 1 \qquad as \ n \to \infty.$$

Note: only the p_i are assumed to vary with n; the quantities t and a_j (with $1 \leqslant j \leqslant t$) are fixed.

Proof. Let $a_0 = n - a_1 - a_2 - \cdots - a_t$. The ratio in the statement of the lemma equals

$$\frac{\binom{n}{a_0,a_1,\ldots,a_t} p_0^{a_0} p_1^{a_1} \cdots p_t^{a_t}}{\prod_{j=1}^{t}\left[\binom{n}{a_j} p_j^{a_j}(1-p_j)^{n-a_j}\right]} \sim \frac{(1-p_1-p_2-\cdots-p_t)^{n-a_1-\cdots-a_t}}{\prod_{j=1}^{t}(1-p_j)^{n-a_j}}$$

$$\sim \frac{\exp\{-(p_1+\cdots+p_t)(n-a_1-\cdots-a_t)\}}{\prod_{j=1}^{t}\exp\{-p_j(n-a_j)\}}$$

$$\sim \frac{\exp\{-(p_1+\cdots+p_t)n\}}{\prod_{j=1}^{t}\exp\{-p_jn\}}$$

$$= 1. \qquad \square$$

The conclusion of the lemma can be restated: the events $N_j = a_j$ (for $1 \leqslant j \leqslant t$) are asymptotically independent. Furthermore, it follows from the lemma that

$$\frac{\Pr\{N_1 \geqslant a_1 \wedge N_2 \geqslant a_2 \wedge \cdots \wedge N_t \geqslant a_t\}}{\Pr\{N_1 \geqslant a_1\}\Pr\{N_2 \geqslant a_2\}\cdots\Pr\{N_t \geqslant a_t\}} \to 1 \qquad \text{as } n \to \infty$$

as well.

Next we consider asymptotic expressions for $\Pr\{N_j \geqslant a_j\}$ that are readily derived from basic properties of the binomial distribution.

Lemma 2. *Suppose we perform n (with $n \to \infty$) Bernoulli trials of an experiment, and the probability of success is p with $p \to 0$ as $n \to \infty$. Let N be the number of successes and let a be a fixed nonnegative integer.*

(1) *If $np \to 0$ then $\Pr\{N \geqslant a\} \sim \Pr\{N = a\} \sim (np)^a/a!$.*
(2) *If $a > 0$ and if there is a constant $\varepsilon > 0$ so that $\varepsilon \leqslant np \leqslant 1/\varepsilon$, then there is a constant $\delta > 0$ so that $\delta \leqslant \Pr\{N \geqslant a\} \leqslant 1 - \delta$.*
(3) *If $np \to \infty$ then $\Pr\{N \geqslant a\} \to 1$.*

Proof. In every case we have $\Pr\{N \geqslant a\} \geqslant \Pr\{N = a\}$. For (1) we compute:

$$\Pr\{N \geqslant a\} = \sum_{j=a}^{n} \Pr\{N = j\} = \sum_{j=a}^{n} \binom{n}{j} p^j (1-p)^{n-j}.$$

Note that in the latter sum the ratio of the successive terms is

$$\frac{\binom{n}{j} p^j (1-p)^{n-j}}{\binom{n}{j+1} p^{j+1}(1-p)^{n-j-1}} = \frac{(j+1)(1-p)}{(n-j)p} > \frac{1}{2np} \to \infty.$$

Bounding the sum by a geometric series we have

$$\Pr\{N = a\} \leqslant \Pr\{N \geqslant a\} \leqslant \Pr\{N = a\}\left(\frac{1}{1-2np}\right) \sim \Pr\{N = a\}.$$

For (2) we have $\Pr\{N \geqslant a\} \geqslant \Pr\{N = a\} = \binom{n}{a}p^a(1-p)^{n-a} \geqslant \binom{n}{a}(\varepsilon/n)^a[1-1/(\varepsilon n)]^{n-a} \sim \varepsilon^a e^{-1/\varepsilon}/a!$, which is a positive constant. On the other hand,

$$\Pr\{N \geqslant a\} = 1 - \sum_{j=0}^{a-1}\Pr\{N = j\} \leqslant 1 - \sum_{j=0}^{a-1}\binom{n}{j}\left(\frac{\varepsilon}{n}\right)^j\left[1-\frac{1}{\varepsilon n}\right]^{n-j} \sim 1 - \sum_{j=0}^{a-1}\frac{\varepsilon^j}{e^{1/\varepsilon}j!},$$

which is strictly less than 1.

Finally, for (3) we note that for fixed j we have $\Pr\{N = j\} = \binom{n}{j}p^j(1-p)^{n-j} \leqslant (np)^j e^{-np} \to 0$ since $np \to \infty$. Hence,

$$\Pr\{N \geqslant a\} = 1 - \sum_{j=0}^{a-1} \Pr\{N = j\} = 1 - o(1). \qquad \square$$

2. Subgraph thresholds

2.1. Thresholds

In this paper we show that for all fixed graphs H there is the 'birth' threshold $\tau_1(H)$ such that, if $p \ll \tau_1(H)$, then with high probability $G(n, m, p)$ does not contain H as a subgraph, while for $p \gg \tau_1(H)$, H is with high probability a subgraph of our random graph. With induced subgraphs there is more to the story. If H is any fixed graph, then the 'birth' threshold for H being an induced subgraph of $G(n, m, p)$ coincides with the 'birth' threshold for H as a subgraph. However the property 'H is an induced subgraph of G' is not monotone; hence, when our random graph becomes dense enough, H will disappear from it. Therefore in this case there are two thresholds, $\tau_1(H)$ and $\tau_2(H)$, associated with H. If $p \ll \tau_1(H)$ or $p \gg \tau_2(H)$, then with high probability $H \not\leqslant G$. However, if $\tau_1(H) \ll p \ll \tau_2(H)$, then with high probability $H \leqslant G$.

Let us first introduce basic notions and a notation used in the paper. Let H be any fixed graph. A *clique cover* of a graph H is a collection of vertex sets such that each induces a complete subgraph (*clique*) of H and, for every edge vw of H, v and w are together in at least one common member of the collection. In other words, the cliques induced by the vertex sets exactly cover the edges of H. We say that $\mathscr{C}$ is *reducible* if, for some $C \in \mathscr{C}$, the edges induced by C are contained in the union of the edges induced by $\mathscr{C}$; otherwise $\mathscr{C}$ is *irreducible*.

If $\mathscr{C} = \{C_1, C_2, \ldots, C_k\}$ is a particular clique cover of H (with $|C_i| \geqslant 1$ for all $i = 1, 2, \ldots, k$) then $|\mathscr{C}|$ denotes the number of cliques in $\mathscr{C}$, $\sum \mathscr{C}$ the sum of clique sizes in $\mathscr{C}$, and $\mathscr{C}'$ stands for $\{C \in \mathscr{C} : |C| > 1\}$.

Furthermore, for $S \subset V(H)$, define *restricted clique covers* as follows:

$$\mathscr{C}[S] := \{C_i \cup S : |C_i \cup S| \geqslant 1, i = 1, 2, \ldots, k\},$$

that is, the clique cover of S that results from restricting the cliques of $\mathscr{C}$ to the vertices that are in S, and

$$\mathscr{C}'[S] := \{C_i \cup S : |C_i \cup S| \geqslant 2, i = 1, 2, \ldots, k\},$$

that is, the clique cover of S that results from restricting the cliques of $\mathscr{C}$ to the vertices that are in S, ignoring all resulting cliques of size 1.

Using these restricted clique covers, let us define:

$$\tau(H, \mathscr{C}, S) = 1 / \left(n^{|S|/\Sigma\mathscr{C}[S]} m^{|\mathscr{C}[S]|/\Sigma\mathscr{C}[S]} \right),$$

$$\tau'(H, \mathscr{C}, S) = 1 / \left(n^{|S|/\Sigma\mathscr{C}'[S]} m^{|\mathscr{C}'[S]|/\Sigma\mathscr{C}'[S]} \right),$$

$$\tau(H, \mathscr{C}) = \max_S \{\tau(H, \mathscr{C}, S), \tau'(H, \mathscr{C}, S)\},$$

$$\tau_1(H) = \min_{\mathscr{C}} \tau(H, \mathscr{C}),$$

where $\mathscr{C}$ is a clique cover of H, and S is a non-empty subset of $V(H)$. When $\mathscr{C}'[S]$ is empty, we put the corresponding τ' term equal to 0.

Furthermore, let $d(H) = |E(H)|/|V(H)|$, while $d^*(H) = \max_{L \leqslant H} d(L)$

We are now ready to state our main result, which considers three segments of the random graph evolution: the 'appearance' period, the period in which all small subgraphs have probability 1 of being present as induced subgraphs, and the 'disappearance' period.

Theorem 3. *Let H be a fixed graph.*

(a) *Suppose $mp^2 \to 0$. Then*

$$\lim_{n \to \infty} \Pr(H \leqslant G(n, m, p)) = \begin{cases} 0 & \text{if} \quad p/\tau_1(H) \to 0, \\ 1 & \text{if} \quad p/\tau_1(H) \to \infty. \end{cases}$$

(b) *Suppose $\epsilon \leqslant mp^2 \leqslant 1/\epsilon$. Then*

$$\lim_{n \to \infty} \Pr(H \leqslant G(n, m, p)) = 1.$$

(c) *Suppose $p = \sqrt{\dfrac{\log n + \omega_n}{d^*(\bar{H})m}}$ and $mp^2 \to \infty$. Then*

$$\lim_{n \to \infty} \Pr(H \leqslant G(n, m, p)) = \begin{cases} 1 & \text{if} \quad \omega_n \to +\infty, \\ 0 & \text{if} \quad \omega_n \to -\infty. \end{cases}$$

Proof. Let $X(H)$ denote the number of copies of H in $G = G(n, m, p)$. Now, if $E(X(H)) \to 0$ as $n \to \infty$, it follows from Markov's inequality that $\Pr\{H \leqslant G\} \to 0$. Furthermore, if L is an induced subgraph of H, and $E(X(L)) \to 0$ as $n \to \infty$, it also follows that $\Pr\{H \leqslant G\} \to 0$. This is the exact same situation as in Erdős–Rényi random graphs.

On the other hand, suppose $E(X(L)) \to \infty$ for all induced subgraphs $L \leqslant H$. In the Erdős–Rényi model this is sufficient to conclude that $H \leqslant G$ with high probability. However, in our model this is not sufficient. Thus the expected number of copies of H in G is not the full story. Nonetheless, it is the beginning of the story, so we concentrate on how to compute it.

Let $\pi(H)$ denote the probability that H is induced on vertices 1 through h *in that order*, that is, the identity map is an isomorphism of H onto the first h vertices of G. Thus, the expected number of copies of H in G is

$$E(X(H)) = \binom{n}{h} \frac{h!}{|\mathrm{aut}(H)|} \pi(H)$$

and it only remains to compute $\pi(H)$.

Let us refine our $X(H)$ notation. Given a clique cover $\mathscr{C}$ of H, let $X(H, \mathscr{C})$ denote the number of copies of H induced in G that are represented by clique cover $\mathscr{C}$. Likewise, let $\pi(H, \mathscr{C})$ denote the probability that H is induced on the first h vertices (in order) of G with clique cover $\mathscr{C}$.

We have now reduced our problem to computing $\pi(H,\mathscr{C})$ We shall show first that, if $mp^2 \to 0$ and $\mathscr{C}$ is a clique cover of a graph H on h vertices, then

$$\pi(H,\mathscr{C}) \begin{cases} \sim m^{|\mathscr{C}|}p^{\Sigma\mathscr{C}}, & mp \to 0, \quad \text{or} \\ \asymp m^{|\mathscr{C}'|}p^{\Sigma\mathscr{C}'}, & mp \geqslant \varepsilon > 0. \end{cases}$$

Hence it follows that

$$EX(H,\mathscr{C}) \asymp \begin{cases} n^h m^{|\mathscr{C}|}p^{\Sigma\mathscr{C}}, & mp \to 0, \quad \text{or} \\ n^h m^{|\mathscr{C}'|}p^{\Sigma\mathscr{C}'}, & mp \geqslant \varepsilon > 0. \end{cases}$$

Let $\mathscr{C} = \{C_1, \ldots, C_t\}$ be a clique cover of a fixed graph H on h vertices. Consider the h rows of $R(n,m,p)$ corresponding to H. The columns in these rows must correspond to cliques in $\mathscr{C}$ (or else contain at most one 1). Thus there are t kinds of columns that are *mandatory* and, say, s kinds of columns that are *forbidden*. The probability that a particular column corresponds to a mandatory clique C_i is $p^{|C_i|}(1-p)^{h-|C_i|} \sim p^{|C_i|}$. Let $N_1, \ldots, N_t$ denote the number of columns corresponding to the cliques in $\mathscr{C}$ and let $N_{t+1}, \ldots, N_{t+s}$ denote the number of columns of the forbidden types. Thus

$$\pi(H,\mathscr{C}) = \Pr\{N_1 > 0 \wedge \cdots N_t > 0 \wedge N_{t+1} = 0 \wedge \cdots \wedge N_{s+t} = 0\},$$

which, by Lemma 1, is asymptotic to

$$\Pr\{N_1 > 0\} \cdots \Pr\{N_t > 0\} \Pr\{N_{t+1} = 0\} \cdots \Pr\{N_{s+t} = 0\}.$$

Now we apply Lemma 2 to the first t terms. For $1 \leqslant i \leqslant t$ we have

$$\Pr\{N_i > 0\} \begin{cases} \sim mp^{|C_i|}, & |C_i| \geqslant 2, \text{ or } |C_i| = 1 \text{ and } mp \to 0, \\ \geqslant \delta, & |C_i| = 1 \text{ and } mp \geqslant \varepsilon, \end{cases}$$

where δ and ε are positive constants.

Next, for $t+1 \leqslant i \leqslant t+s$, the kind of column we are forbidding has $a \geqslant 2$ ones. Thus,

$$\Pr\{N_i = 0\} = (1 - p^a q^{h-a})^m \sim \exp\{-mp^a q^{h-a}\} \sim 1$$

since $mp^2 \to 0$ (and $a \geqslant 2$).

Combining these results we get either

$$\pi(H,\mathscr{C}) \sim m^{|\mathscr{C}|}p^{\Sigma\mathscr{C}} \qquad \text{provided } mp \to 0,$$

or else,

$$\pi(H,\mathscr{C}) \asymp m^{|\mathscr{C}'|}p^{\Sigma\mathscr{C}'} \qquad \text{provided } mp \geqslant \varepsilon > 0.$$

We now claim that reducible clique covers are less likely to occur than irreducible ones. Suppose $mp^2 \to 0$, and $\mathscr{C}$ is a reducible clique cover of H. Thus there is a $C \in \mathscr{C}$ so that $\mathscr{C}^* = \mathscr{C} - \{C\}$ is also a clique cover of H. Now if C is a 1-clique we clearly have $\pi(H,\mathscr{C}) \leqslant \pi(H,\mathscr{C}^*)$ since C is *permitted* in the $\mathscr{C}^*$ representation, but *required* in the $\mathscr{C}$ representation. Otherwise, $|C| \geqslant 2$ and we have $\pi(H,\mathscr{C}) \leqslant mp^2\pi(H,\mathscr{C}^*) \ll \pi(H,\mathscr{C}^*)$ (since $mp^2 \to 0$).

To prove part (a) of our theorem, assume as before that $\mathscr{C}$ is an irreducible clique cover of H and let S be a non-empty subset of $V(H)$. Then the *restriction* of $\mathscr{C}$ to S is the multiset $\mathscr{C}[S] = \{C \cap S \neq \emptyset : C \in \mathscr{C}\}$.

We then let $\pi(H, \mathscr{C}, S)$ be the probability that a fixed set of $|S|$ rows generates $\mathscr{C}[S]$, i.e., that for each $C \in \mathscr{C}[S]$ there is a separate column in the rows corresponding to S with 1s exactly for the elements of C.

Let $X(H, \mathscr{C}, S)$ be the number of subsets of rows of $R(n, m, p)$ that generate $\mathscr{C}[S]$. Thus $EX(H, \mathscr{C}, S) \asymp n^{|S|} \pi(H, \mathscr{C}, S)$. Therefore

$$EX(H, \mathscr{C}, S) \asymp \begin{cases} x = n^{|S|} m^{|\mathscr{C}[S]|} p^{\Sigma \mathscr{C}[S]}, & mp \to 0, \quad \text{and} \\ x' = n^{|S|} m^{|\mathscr{C}'[S]|} p^{\Sigma \mathscr{C}'[S]}, & mp \not\to 0. \end{cases}$$

where $\mathscr{C}'[S] = \{C \cap S : C \in \mathscr{C}, |C \cap S| > 1\}$.

Our next step in deriving a formula for $\tau_1(H)$ – the appearance threshold for H – is to show that the following statements hold. First, if for some $S \subseteq V(H)$ we have

$$n^{|S|} m^{|\mathscr{C}[S]|} p^{\Sigma \mathscr{C}[S]} \to 0 \text{ or } n^{|S|} m^{|\mathscr{C}'[S]|} p^{\Sigma \mathscr{C}'[S]} \to 0$$

as $n \to \infty$, then we also have $\Pr\{X(H, \mathscr{C}) > 0\} \to 0$. Second, if for all $S \subseteq V(H)$ we have

$$n^{|S|} m^{|\mathscr{C}[S]|} p^{\Sigma \mathscr{C}[S]} \to \infty \text{ and } n^{|S|} m^{|\mathscr{C}'[S]|} p^{\Sigma \mathscr{C}'[S]} \to \infty$$

as $n \to \infty$, then we also have $\Pr\{X(H, \mathscr{C}) > 0\} \to 1$.

Since $X(H, \mathscr{C}, S) = 0 \implies X(H, \mathscr{C}) = 0$, it is enough to show that $EX(H, \mathscr{C}, S) \to 0$ for some $S \subseteq V(H)$. We have four possible cases to consider, depending on whether an x or x' tends to 0, and depending on whether or not mp tends to 0.

Observe that x and x' differ by a power of mp, namely $x = (mp)^{\ell} x'$ for some integer $\ell \geqslant 0$ (where ℓ is the number of 1-cliques in $\mathscr{C}[S]$).

Suppose first that $mp \to 0$. If $x \to 0$ (for some S) then, since $x \asymp EX(H, \mathscr{C}, S)$, we are done. Otherwise, if some $x' \to 0$, then, since $x = (mp)^{\ell} x'$ we also have $x \to 0$, and, again, we are done. On the other hand, suppose $mp \geqslant \varepsilon > 0$. If some $x' \to 0$ then, since $x' \asymp EX(H, \mathscr{C}, S)$, we are done. Otherwise, if some $x \to 0$, then since $x' = x/(mp)^{\ell} \leqslant x/\varepsilon^{\ell}$ we must have $x' \to 0$, and again we are done. Thus, if any expression of the form $n^{|S|} m^{|\mathscr{C}[S]|} p^{\Sigma \mathscr{C}[S]}$ or $n^{|S|} m^{|\mathscr{C}'[S]|} p^{\Sigma \mathscr{C}'[S]}$ tends to 0, $X(H, \mathscr{C}) = 0$, almost surely.

Now, suppose that for all S we have $n^{|S|} m^{|\mathscr{C}[S]|} p^{\Sigma \mathscr{C}[S]}$ and $n^{|S|} m^{|\mathscr{C}'[S]|} p^{\Sigma \mathscr{C}'[S]}$ tending to infinity. Let $\mu = EX(H, \mathscr{C}) = EX(H, \mathscr{C}, V(H))$, so $\mu \to \infty$. We show that $\Pr\{X(H, \mathscr{C}) > 0\} \to 1$ by the second moment method. Write

$$E[X(H, \mathscr{C})^2] = \sum_A \sum_B E[Z_A Z_B]$$

where the sums are over all h element subsets of $[n]$ and Z_A is 1 when the rows of $R(n, m, p)$ corresponding to A generate a copy of H with clique cover $\mathscr{C}$ and 0 otherwise. When $A \cap B = \emptyset$, note that Z_A and Z_B are independent; there are $\binom{n}{h}\binom{n-h}{h}$ such pairs (A, B). Thus,

$$E[X(H, \mathscr{C})^2] \sim \mu^2 + \sum_{A \cap B \neq \emptyset} E[Z_A Z_B].$$

We wish to show that $E[X(H, \mathscr{C})^2] \sim \mu^2$, so we need to show that the above summation is $o(\mu^2)$. There are $O(n^{2h-s})$ pairs (A, B) for which $|A \cap B| = s$, so it is enough to prove that when $|A \cap B| = s$ we have

$$(n^{2h-s} E[Z_A Z_B])/\mu^2 \to 0. \tag{*}$$

Let $\mathscr{C}_A$ be the clique cover $\mathscr{C}$ in terms of the specific assignment of cliques to labelled vertex sets as indicated by Z_A (since Z_A is associated with a particular order for laying down the clique cover on a set of h vertices). Define $\mathscr{C}_B$ similarly.

Let $\mathscr{C}^*$ be the cover of $G[A \cup B]$ generated by taking the union of the collections $\mathscr{C}_A$ and $\mathscr{C}_B$. By this definition, some edge sets of $A \cap B$ may get covered more than once, for example, if they are covered by different cliques in $\mathscr{C}_A$ and $\mathscr{C}_B$. The other sets will get covered exactly once.

For the purpose of comparing the numerator and denominator of the ratio

$$(n^{2h-s}\mathbf{E}[Z_A Z_B])/\mu^2,$$

think of writing the asymptotic expression for μ^2 as $\mu_A\mu_B$ (with $\mu_A = \mu_B$), where the clique terms in the two products μ_A and μ_B are ordered in the same way as the cliques in $\mathscr{C}_A$ and $\mathscr{C}_B$.

By definition of a union, there is a clique in $\mathscr{C}^*$ for every clique of $\mathscr{C}_A$, yielding terms of the form $mp^{|C|}$ in both the numerator and denominator for each such clique $|C|$. Thus the final product of clique probabilities for A in the numerator cancels with the clique probability part of μ_A in the denominator.

And what about those terms stemming from cliques in $\mathscr{C}_B$? For a clique C in $\mathscr{C}^*$ that consists only of vertices from B (i.e., $C \cap A = \emptyset$), the numerator and μ_B will both have terms of the form $mp^{|C|}$, which will cancel each other. The only remaining type of clique probability term in the numerator will be for those cliques of $\mathscr{C}^*$ that contain vertices of $A \cap B$ but are not cliques of $\mathscr{C}_A$ (if they were in $\mathscr{C}_A$ they would already have been cancelled out with the cliques of A). Since these are additional cliques of $\mathscr{C}_B$, they match terms in the denominator's μ_B, and cancel with them.

The left-over clique probability terms are then all in the μ_B part of the denominator, and correspond to cliques on vertices of $A \cap B$ that are in both covers $\mathscr{C}_A$ and $\mathscr{C}_B$. For this reason, there was only one copy of the term in the numerator originally, and it was cancelled out when $\mathscr{C}_A$ was examined. There was, however, a copy in each of μ_A and μ_B, and the one in μ_B remains.

Letting $C_1, C_2, \ldots, C_b$ refer to the cliques of $\mathscr{C}_B$ generating these remaining terms, the ratio for $(n^{2h-s}\mathbf{E}[Z_A \cap A_B])/\mu^2$ can be simplified to

$$\frac{n^{2h-s}}{n^{2h}}(1)\left(\prod_{i=1}^{b}\frac{1}{mp^{|C_i|}}\right) = \frac{1}{n^s\prod_{i=1}^{b}mp^{|C_i|}}.$$

Each C_i here is a clique of $\mathscr{C}_B$ on $S = (A \cap B) \subseteq B$. So this is a partial set of the cliques in $\mathscr{C}[S]$. By assumption, $n^s m^{|\mathscr{C}[S]|}p^{\Sigma\mathscr{C}[S]} \to \infty$. But this expression can be written as a product of the current term $n^s\prod_{i=1}^{b}mp^{|C_i|}$ with additional terms of the form mp^a, each for some $a \geq 2$, and all of which tend to 0. Thus $n^s\prod_{i=1}^{b}mp^{|C_i|} \gg n^s m^{|\mathscr{C}[S]|}p^{\Sigma\mathscr{C}[S]} \to \infty$, and so $1/(n^s\prod_{i=1}^{b}mp^{|C_i|})$ tends to 0 as $n \to \infty$.

We can now derive a formula for the appearance threshold for H. Check that we have selected $\tau_1(H)$ so that if $p \ll \tau_1(H)$ then for every $\mathscr{C}$ there is an $S \subseteq V(H)$ for which $n^{|S|}m^{|\mathscr{C}[S]|}p^{\Sigma\mathscr{C}[S]}$ or $n^{|S|}m^{|\mathscr{C}'[S]|}p^{\Sigma\mathscr{C}'[S]}$ tends to 0. Conversely, if $p \gg \tau_1(H)$ then there is a $\mathscr{C}$ so that for all $S \subseteq V(H)$ we have $n^{|S|}m^{|\mathscr{C}[S]|}p^{\Sigma\mathscr{C}[S]}$ and $n^{|S|}m^{|\mathscr{C}'[S]|}p^{\Sigma\mathscr{C}'[S]}$ tending to infinity.

Thus we have shown that if H is a fixed graph and $mp^2 \to 0$, then if $p/\tau_1(H) \to 0$ then with high probability H is not an induced subgraph of $G(n, m, p)$ while if $\tau_1(H)/p \to 0$ then with high probability H is an induced subgraph of $G(n, m, p)$.

To prove part (b) of our theorem, that is, to show that when mp^2 is bounded away from 0 then the probability that H is an induced subgraph of $G(n, m, p)$ tends to 1 as $n \to \infty$, let $\mathscr{C} = E(H)$, that is, consider the clique cover of H consisting of all pairs of adjacent vertices in H. Let N_i (for $1 \leqslant i \leqslant |E(H)|$) denote the number of columns corresponding to the ith edge of H. Ordering all other types of columns that have at least two 1s, let N_i for $i > |E(H)|$ denote the number of columns of type i. We want to compute

$$\Pr\left\{ N_1 > 0 \wedge \cdots \wedge N_{|E(H)|} > 0 \wedge N_{|E(H)|+1} = 0 \wedge \cdots \right\},$$

which, by Lemma 1, is asymptotic to

$$\Pr\left\{ N_1 > 0 \right\} \Pr\left\{ N_2 > 0 \right\} \cdots \Pr\left\{ N_{|E(H)|} > 0 \right\} \Pr\left\{ N_{|E(H)|+1} = 0 \right\} \Pr\left\{ N_{|E(H)|+2} = 0 \right\} \cdots.$$

Now, we notice that by Lemma 2 we have, for $1 \leqslant i \leqslant |E(H)|$,

$$\Pr\left\{ N_i > 0 \right\} \in [\delta, 1 - \delta]$$

for some positive constant δ. For $i > |E(H)|$ we have $\Pr\left\{ N_i = 0 \right\} \geqslant \delta$. Thus $\pi(H, \mathscr{C}) \geqslant \delta'$, where δ' is a positive constant (a fixed power of δ).

We can decompose the n vertices of $G(n, m, p)$ into $k = \lfloor n/h \rfloor$ pairwise disjoint sets of h vertices $S_1, S_2, \ldots, S_k$ (plus, perhaps, a few left-overs) and we note that the probability of having a copy of H induced on any of them is at least δ' and these events are mutually independent. Since $k \to \infty$, we have $H \leqslant G(n, m, p)$ with probability tending to 1.

Finally, to prove the last statement (c) of our main theorem, consider the situation when $mp^2 \to \infty$. Then notice that the probability ρ that two vertices v and w are *not* adjacent:

$$\rho = \Pr\left\{ vw \notin E(G) \right\} = (1 - p^2)^m \sim e^{-mp^2},$$

which tends to 0. Thus it is the non-edges that are difficult to form.

Let us see that we may, without any loss of generality, assume that $mp^3 \to 0$. Let Z denote the number of non-edges in $G(n, m, p)$. Thus $EZ = \binom{n}{2}\rho \sim \frac{1}{2}n^2 e^{-mp^2}$. Thus if $p = \sqrt{(2\log n + \omega_n)/m}$ (where $\omega_n \to \infty$) then $EZ \to 0$ and with high probability $G(n, m, p)$ is complete. In this case $mp^3 = (2\log n + \omega_n)^{3/2}/\sqrt{m}$, which tends to 0 (unless ω_n is very large).

As before, let $X(H)$ denote the number of copies of H in $G(n, m, p)$. Let $X_2(H) = X(H, E(H))$, that is, the number of copies of H appearing in $G(n, m, p)$ with clique cover consisting of exactly the edges of H. (The subscript '2' refers to the fact that all cliques in this cover have size 2.)

First, we shall prove that if $mp^2 \to \infty$, $mp^3 \to 0$ and H is a fixed graph, then

$$EX(H) \sim EX_2(H) \sim \binom{n}{h}\rho^{|E(\overline{H})|}$$

where $\rho \sim e^{-mp^2}$. To see this let us first consider the clique cover $\mathscr{C} = E(H)$ on a fixed set of h vertices. Let N_i (with $1 \leqslant i \leqslant |E(H)|$) denote the number of columns corresponding to the ith clique (edge) of $\mathscr{C}$. For $|E(H)| < i \leqslant \binom{h}{2}$, let N_i denote the number of columns that generate an edge on the respective *non-edge* pair of vertices of H (columns with 2 ones that are located in particular rows not corresponding to end-points of an edge of H). For $i > \binom{h}{2}$, the N_i denote the number of cliques on 3 or more of the h vertices. We want $N_i > 0$ for $1 \leqslant i \leqslant |E(H)|$ and $N_i = 0$ for $i > |E(H)|$.

Applying Lemmas 1 and 2, we have

$$EX_2(H) \sim \binom{n}{h} \rho^{\binom{h}{2} - |E(H)|} = \binom{n}{h} \rho^{|E(\overline{H})|}.$$

Now if $\mathscr{C}$ is any *other* clique cover of H, then we still have $E(\overline{H}) \cap \mathscr{C} = \emptyset$, but we have some replacement clique(s) in $\mathscr{C}$. For each one that has 3 or more vertices, a term of the form mp^c (with $c \geqslant 3$) replaces mp^2 terms in our expression $EX(H, \mathscr{C})$ and, since $mp^3 \to 0$, we have $EX(H, \mathscr{C}) \ll EX_2(H)$. Thus we can restrict our attention to clique covers of H consisting purely of cliques of size 2, and if $EX_2(H) \to 0$, then with high probability H is not an induced subgraph of $G(n, m, p)$. Further, if $EX_2(L) \to 0$ for any induced subgraph of H, then, again, H is with high probability not an induced subgraph.

Now suppose that $p = \sqrt{(\log n + \omega_n)/(d^*(\overline{H})m)}$. Let L be such that its complement $\overline{L}$ satisfies $\overline{L} \leqslant \overline{H}$ and $d(\overline{L}) = d^*(\overline{H})$, so $d^*(\overline{H}) = E(\overline{L})/V(L)$. Put $\ell = |V(L)|$. Then

$$EX_2(L) \asymp n^\ell \rho^{|E(\overline{L})|} \sim n^\ell \exp\{-mp^2|E(\overline{L})|\} = n^\ell n^{-|E(\overline{L})|/d^*(\overline{H})} e^{-\omega_n E(\overline{L})/d^*(\overline{H})}$$
$$= o(1) n^\ell n^{-\ell} \to 0.$$

Thus, with high probability $H \not\leqslant G(n, m, p)$.

On the other hand, suppose $p = \sqrt{(\log n - \omega_n)/(d^*(\overline{H})m)}$. In this case $EX_2(L) \to \infty$ for all $L \leqslant H$. Put $\mu = EX_2(H)$. To use the second moment method, we compute $E[X_2(H)^2]$ and work to show that it is asymptotic to μ^2. We can decompose $E[X_2(H)^2]$ as

$$E[X_2(H)^2] = \sum_{A,B} E[Z_A Z_B],$$

where the sum is over all $h = |V(H)|$ element subsets of $V(G)$, and Z_A is an indicator random variable that is 1 exactly when $G[A]$ is a copy of H. When $A \cap B = \emptyset$ then Z_A and Z_B are independent. There are $\binom{n}{h}\binom{n-h}{h} \sim \binom{n}{h}^2$ such terms, so $\sum_{A,B:|A\cap B|=\emptyset} E[Z_A Z_B] \sim \binom{n}{h}^2 E[Z_A]^2 = \mu^2$. Thus, it remains to show that

$$\sum_{A,B:|A\cap B|=\ell} E[Z_A Z_B] = o(\mu^2),$$

where ℓ is a fixed integer with $1 \leqslant \ell \leqslant h$. There are some $n^{2h-\ell}$ such pairs of sets A and B. The probability we have $Z_A Z_B = 1$ is just ρ^k, where k is the number of non-edges in $G[A \cup B]$. Note that $k = 2|E(\overline{H})| - |E(\overline{L})|$ where $L = G[A \cap B]$. Thus, comparing to $\mu^2 \asymp n^{2h} \rho^{2|E(\overline{H})|}$ we have

$$\frac{n^{2h-\ell} \rho^k}{n^{2h} \rho^{2|E(\overline{H})|}} = \frac{1}{n^\ell \rho^{|E(\overline{L})|}} \sim \frac{1}{EX_2(L)} \to 0,$$

as we claimed. $\square$

The following result follows from our main theorem.

Corollary 4 (Large α). *For a fixed graph H, there is an $\alpha^* > 0$ such that*

$$\tau_1(H) = 1/(n^{1/2d^*(H)}m^{1/2}) \text{ for all } \alpha \geqslant \alpha^*,$$

where $d^*(H) = \max_{L \leqslant H} |E(L)|/|V(L)|$

Proof. This threshold arises from taking $\mathscr{C} = E(H)$ (*i.e.*, letting $\mathscr{C}$ be the clique cover consisting of exactly the edges of H), and letting S be a set of vertices that induces a subgraph of maximum density in H. What is the asymptotic probability of an edge between two vertices when p is at this threshold? We have $mp^2 = m(\tau_1(H))^2 = n^{-1/d^*(H)}$, which is exactly the probability threshold for the appearance of H as a subgraph in the Erdős–Rényi model $G(n, p)$.

The proof of this threshold begins by showing that, for any particular $\mathscr{C}$,

$$\max_S \{\tau(H, \mathscr{C}, S), \tau'(H, \mathscr{C}, S)\}$$

may be found by considering only τ' for each S (not τ), as long as α is large enough. When $\mathscr{C} = E(H)$, $\max_S \{\tau'(H, \mathscr{C}, S)\}$ occurs when $S = V(L)$ for $L \leqslant H : |E(H)|/|V(H)| = d^*(H)$, giving $\tau'(H, \mathscr{C}) = 1/(n^{1/2d^*(H)}m^{1/2})$. It can be shown (see [19]) that for any other clique cover there is some choice of S that gives a value of $\tau'(H, \mathscr{C}, S)$ that is greater than the one above when α is large, and so the maximum over $S \subseteq V$ for that clique cover will be at least that big. The justification that there is some such choice of S for each $\tilde{\mathscr{C}} \neq E(H)$ relies on the definition of τ' and the fact that, since $\tilde{\mathscr{C}} \neq E(H)$, there is at least one S for which $|\mathscr{C}[S]|/\Sigma\tilde{\mathscr{C}}[S] < 1/2$. Thus, $\min_{\mathscr{C}} \max_S \{\tau(H, \mathscr{C}, S), \tau'(H, \mathscr{C}, S)\} = 1/(n^{1/2d^*(H)}m^{1/2})$. $\square$

Finally, let us notice that statement (a) of Theorem 3 can also be deduced from a bipartite version of the main result of paper [10]. Indeed, as we pointed out in the Introduction, the binomial bipartite model has a simple equivalence to the random matrix model, and can thus generate graphs from $G(n, m, p)$ according to the transformation from $\mathscr{R}(n, m, p)$ to $\mathscr{G}(n, m, p)$. This equivalence provides a useful relationship between subgraphs of $B(n, m, p)$ and clique covers for subgraphs of $G(n, m, p)$. As a result of this relationship several other results of [10], dealing for example with the distribution of the number of certain classes of strictly balanced subgraphs, can be applied to the study of subgraphs of $G(n, m, p)$ (see [19] for details).

2.2. Example

Let H be a fixed graph on h vertices, such as the graph in Figure 1, and let $X(H)$ denote the number of copies of H induced in $G = G(n, m, p)$.

Now there are a number of possible ways in which H can appear in G. Vertices 1, 2 and 3 form a K_3. In the representation of G (*i.e.*, the matrix R) there may be a column in which vertices 1, 2 and 3 have a common 1, or there may be three columns representing separate edges between the pairs. Thus the first four rows of R look like one of the

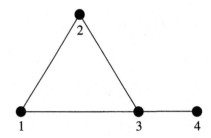

Figure 1   The graph H.

following:

$$
\begin{bmatrix}
\cdots & 1 & \cdots & 0 & \cdots \\
\cdots & 1 & \cdots & 0 & \cdots \\
\cdots & 1 & \cdots & 1 & \cdots \\
\cdots & 0 & \cdots & 1 & \cdots
\end{bmatrix}
\quad \text{or} \quad
\begin{bmatrix}
\cdots & 1 & \cdots & 0 & \cdots & 1 & \cdots & 0 & \cdots \\
\cdots & 1 & \cdots & 1 & \cdots & 0 & \cdots & 0 & \cdots \\
\cdots & 0 & \cdots & 1 & \cdots & 1 & \cdots & 1 & \cdots \\
\cdots & 0 & \cdots & 0 & \cdots & 0 & \cdots & 1 & \cdots
\end{bmatrix}.
$$

These possibilities correspond exactly to two possible clique covers of the example graph H: $\mathscr{C}_1 = \{\{1,2,3\},\{3,4\}\}$ and $\mathscr{C}_2 = \{\{1,2\},\{2,3\},\{1,3\},\{3,4\}\}$.

Given a clique cover $\mathscr{C}$ of H, let $X(H,\mathscr{C})$ denote the number of copies of H induced in G that are represented by clique cover $\mathscr{C}$. Likewise, let $\pi(H,\mathscr{C})$ denote the probability that H is induced on the first h vertices (in order) of G with clique cover $\mathscr{C}$.

Let us first consider the clique cover $\mathscr{C}_1 = \{\{1,2,3\},\{3,4\}\}$. How could the first four rows of R fail to create this clique cover? We might

(i) be missing a column of the form $[1,1,1,0]^T$,

(ii) be missing a column of the form $[0,0,1,1]^T$, or

(iii) have a 'bad' column such as $[1,0,0,1]^T$ (which would create an unwanted edge between vertices 1 and 4).

We have

$$
\pi(H,\mathscr{C}) = \sum_{\substack{i,j>0 \\ i+j \leq m}} \binom{m}{i,j,m-i-j} [p^3 q]^i \, [p^2 q^2]^j \, [q^4 + 4pq^3]^{m-i-j}, \tag{$**$}
$$

where $q = 1 - p$. Note that i records the number of columns of type $[1,1,1,0]^T$ and j records the number of columns of type $[0,0,1,1]^T$. The $[q^4 + 4pq^3]^{m-i-j}$ term is for the probability that the remaining $m-i-j$ columns are acceptable, that is, have at most one 1. The right-hand side of $(**)$ simplifies to

$$
(p^3 q + p^2 q^2 + 4pq^3 + q^4)^m - (p^3 q + 4pq^3 + q^4)^m - (p^2 q^2 + 4pq^3 + q^4)^m + (4pq^3 + q^4)^m
$$

which, provided $mp^2 \to 0$, is asymptotic to $m^2 p^5$. By a similar method, one can show that $\pi(H,\mathscr{C}_2) \sim m^4 p^8$.

It is important to note that 1-cliques play a special, exceptional role. Neither $\mathscr{C}_1$ nor $\mathscr{C}_2$ *require* there to be any 1-cliques, but we *allow* them to appear in the representation. Thus we define $\pi(H,\mathscr{C})$ (respectively, $X(H,\mathscr{C})$) to *allow* a 1-clique if it is *not* listed in $\mathscr{C}$, but to *require* a 1-clique if it *is* listed in $\mathscr{C}$. For example, if $mp \to 0$, we find that

$$\pi(H, \mathscr{C}_1) = m^2 p^5,$$
$$\pi(H, \mathscr{C}_1 \cup \{\{1\}\}) = m^3 p^6.$$

Note that $\mathscr{C}_1$ and $\mathscr{C}_2$ are *not* the only clique covers of H: there are a number of other possibilities. We might add to either $\mathscr{C}_1$ or $\mathscr{C}_2$ some singleton cliques (thereby changing their existence from *permitted* to *required*). Or we might add $\{1, 2\}$ to $\mathscr{C}_1$ (changing the status of the clique $\{1, 2\}$ from absent to present). Indeed, $\mathscr{C}_1 \cup \mathscr{C}_2$ is a clique cover as well. However, $\mathscr{C}_1$ and $\mathscr{C}_2$ are the only *irreducible* clique covers of H.

We have

$$EX(H, \mathscr{C}_1) \asymp n^4 m^2 p^5 \quad \text{and} \quad EX(H, \mathscr{C}_2) \asymp n^4 m^4 p^8.$$

Thus, if

$$p \ll 1/\left(n^{4/5} m^{2/5}\right) \quad \text{and} \quad p \ll 1/\left(n^{1/2} m^{1/2}\right),$$

then $E[X(H, \mathscr{C}_1)]$ and $E[X(H, \mathscr{C}_2)]$ tend to 0 as $n \to \infty$. Thus, if we let

$$p = \frac{1}{\omega_n} \min\left\{n^{-4/5} m^{-2/5}, n^{-1/2} m^{-1/2}\right\} = \begin{cases} 1/(\omega_n n^{4/5} m^{2/5}) & \alpha \leqslant 3, \\ 1/(\omega_n n^{1/2} m^{1/2}) & \alpha \geqslant 3, \end{cases}$$

then with high probability $H \not\leqslant G(n, m, p)$.

Now, if any of H's induced subgraphs L has $E[X(L)] \to 0$, then we may also conclude that $\Pr\{H \leqslant G\} \to 0$. The probabilities derived from this, however, do not determine the threshold for H appearing as an induced subgraph of G.

For example, taking $\alpha = 0.2$ and $p = 1/(n^{0.6} m) \gg 1/(n^{4/5} m^{2/5})$, we have $EX(H) \geqslant EX(H, \mathscr{C}_1) \asymp n^4 m^2 p^5 = n^{0.4} \to \infty$. Likewise, one can check that $EX(L) \to \infty$ for all $L \leqslant H$. However, we claim that $\Pr\{H \leqslant G\} \to 0$. Why? Consider vertex 3. In its row there must be at least two 1s: one 1 for the connection to vertex 4, and a second (and possibly third) 1 for the connection to vertices 1 and 2. Let Y denote the number of rows of G's representation matrix with at least two 1s. The expected number of 1s in a given row is mp, which is, in this case, $n^{-0.6}$. The probability that there are two (or more) ones in a given row is asymptotically $(mp)^2 = n^{-1.2}$. Thus $EY \asymp n^{-0.2}$, and therefore with high probability there are no such rows in G's representation. With high probability it is impossible for $H \leqslant G$.

Thus it is not enough that $E(X(L)) \to \infty$ for all $L \leqslant H$ to ensure $H \leqslant G$ with high probability.

What has gone 'wrong'? If we consider the clique cover $\mathscr{C}_1$ or $\mathscr{C}_2$ *restricted* to the one element set $S = \{2\}$ we get a *restricted clique cover* of H induced on S. We denote this by $\mathscr{C}_1[S] = \{\{2\}, \{2\}\}$ (notice that this is a multiset). It is necessary that there be columns in the representation matrix of G corresponding to these restricted clique covers.

Consider the graph of Figure 1. It has two (irreducible) clique covers:

$$\mathscr{C}_1 = \{\{1, 2, 3\}, \{3, 4\}\} \quad \text{and} \quad \mathscr{C}_2 = \{\{1, 2\}, \{1, 3\}, \{2, 3\}, \{3, 4\}\}.$$

For each we determine $\mathscr{C}[S]$ and $\mathscr{C}'[S]$ for all non-empty subsets of S and then compute $\tau(H, \mathscr{C}, S)$ and $\tau'(H, \mathscr{C}, S)$. These calculations are collected in Table 1.

To compute $\tau(H, \mathscr{C}_1)$ (respectively, $\tau(H, \mathscr{C}_2)$) we need to find the largest entry in the

Table 1   Computations for $\tau_1(H)$ for the graph in Figure 1.

S	$\mathscr{C}_1[S]$	$\tau(H,\mathscr{C}_1,S)$	$\mathscr{C}_1'[S]$	$\tau'(H,\mathscr{C}_1,S)$
$\{1,2,3,4\}$	$\{123,34\}$	$1/(n^{4/5}m^{2/5})$	$\{123,34\}$	$1/(n^{4/5}m^{2/5})$
$\{1,2,3\}$	$\{123,3\}$	$1/(n^{3/4}m^{1/2})$	$\{123\}$	$1/(nm^{1/3})$
$\{1,2,4\}$	$\{12,4\}$	$1/(nm^{2/3})$	$\{12\}$	$1/(n^{3/2}m^{1/2})$
$\{1,3,4\}$	$\{13,34\}$	$1/(n^{3/4}m^{1/2})$	$\{13,34\}$	$1/(n^{3/4}m^{1/2})$
$\{1,2\}$	$\{12\}$	$1/(nm^{1/2})$	$\{12\}$	$1/(nm^{1/2})$
$\{1,3\}$	$\{13,3\}$	$1/(n^{2/3}m^{2/3})$	$\{13\}$	$1/(nm^{1/2})$
$\{1,4\}$	$\{1,4\}$	$1/(nm)$	$\emptyset$	—
$\{3,4\}$	$\{3,34\}$	$1/(n^{2/3}m^{2/3})$	$\{34\}$	$1/(nm^{1/2})$
$\{1\}$	$\{1\}$	$1/(nm)$	$\emptyset$	—
$\{3\}$	$\{3,3\}$	$1/(n^{1/2}m)$	$\emptyset$	—
$\{4\}$	$\{4\}$	$1/(nm)$	$\emptyset$	—

S	$\mathscr{C}_2[S]$	$\tau(H,\mathscr{C}_2,S)$	$\mathscr{C}_2'[S]$	$\tau'(H,\mathscr{C}_2,S)$
$\{1,2,3,4\}$	$\{12,13,23,34\}$	$1/(n^{1/2}m^{1/2})$	$\{12,13,23,34\}$	$1/(n^{1/2}m^{1/2})$
$\{1,2,3\}$	$\{12,13,23,3\}$	$1/(n^{3/7}m^{4/7})$	$\{12,13,23\}$	$1/(n^{1/2}m^{1/2})$
$\{1,2,4\}$	$\{12,1,2,4\}$	$1/(n^{3/5}m^{4/5})$	$\{12\}$	$1/(n^{3/2}m^{1/2})$
$\{1,3,4\}$	$\{1,13,3,34\}$	$1/(n^{1/2}m^{2/3})$	$\{13,34\}$	$1/(n^{3/4}m^{1/2})$
$\{1,2\}$	$\{12,1,2\}$	$1/(n^{1/2}m^{3/4})$	$\{12\}$	$1/(nm^{1/2})$
$\{1,3\}$	$\{1,13,3,3\}$	$1/(n^{2/5}m^{4/5})$	$\{13\}$	$1/(nm^{1/2})$
$\{1,4\}$	$\{1,1,4\}$	$1/(n^{2/3}m)$	$\emptyset$	—
$\{3,4\}$	$\{3,3,34\}$	$1/(n^{1/2}m^{3/4})$	$\{34\}$	$1/(nm^{1/2})$
$\{1\}$	$\{1,1\}$	$1/(n^{1/2}m)$	$\emptyset$	—
$\{3\}$	$\{3,3,3\}$	$1/(n^{1/3}m)$	$\emptyset$	—
$\{4\}$	$\{4\}$	$1/(nm)$	$\emptyset$	—

upper (respectively, lower) portion of Table 1. Which is largest depends on α, and we achieve the following:

$$\tau(H,\mathscr{C}_1) = \begin{cases} 1/\left(n^{1/2}m\right) & \alpha \leqslant 1/2, \\ 1/\left(n^{4/5}m^{2/5}\right) & 1/2 \leqslant \alpha \leqslant 3, \\ 1/\left(nm^{1/3}\right) & \alpha \geqslant 3, \end{cases}$$

and $$\tau(H,\mathscr{C}_2) = \begin{cases} 1/\left(n^{1/3}m\right) & \alpha \leqslant 2/9, \\ 1/\left(n^{3/7}m^{4/7}\right) & 2/9 \leqslant \alpha \leqslant 1, \\ 1/\left(n^{1/2}m^{1/2}\right) & \alpha \geqslant 1. \end{cases}$$

Check that for $\alpha \leqslant 3$ we have $\tau(H,\mathscr{C}_1) \leqslant \tau(H,\mathscr{C}_2)$, but for $\alpha \geqslant 3$ we have $\tau(H,\mathscr{C}_1) \geqslant \tau(H,\mathscr{C}_2)$. Thus,

$$\tau_1(H) = \begin{cases} 1/\left(n^{1/2}m\right) & \alpha \leqslant 1/2, \\ 1/\left(n^{4/5}m^{2/5}\right) & 1/2 \leqslant \alpha \leqslant 3, \\ 1/\left(n^{1/2}m^{1/2}\right) & \alpha \geqslant 3. \end{cases}$$

From our main theorem it easily follows that the 'death' threshold for H is

$$\tau_2(H) = \sqrt{\frac{3\log n + \omega_n}{2m}}.$$

2.3. The evolutions: thresholds for specific subgraphs

Let us apply our results to some specific subgraphs such as cycles, trees, complete graphs and triangle-free graphs of a given order. In this section we shall present 'birth' thresholds and some 'death' thresholds for induced subgraphs of specified type. The computations can be laborious in some cases. The details of the proof for the complete graph case are provided as an example. For details of the other proofs see [19].

Corollary 5 (Cycles). *If $h > 3$ then $\tau_1(C_h) = 1/\sqrt{nm}$, while $\tau_2(C_h) = \sqrt{\frac{2\ln n}{(h-3)m}}$.*

Corollary 6 (Complete graphs). *For a complete graph K_h, we have*

$$\tau_1(K_h) = \begin{cases} 1/(nm^{1/h}) & \text{for } \alpha \leqslant 2h/(h-1), \quad \text{and} \\ 1/(n^{1/(h-1)}m^{1/2}) & \text{for } \alpha \geqslant 2h/(h-1). \end{cases}$$

The above threshold expressions arise from the expectations under the clique covers $\{V(H)\}$ (consisting of a single K_h) and $E(H)$ (the edge cover), respectively. Their validity is proved by a series of three claims. The proofs sometimes use V to refer to $V(H)$.

Claim 1. *For the single-clique cover $\{V(H)\}$,*

$$\max_{S}\{\tau(K_h, \{V(H)\}, S), \tau'(K_h, \{V(H)\}, S)\} = \tau(K_h, \{V(H)\}, V).$$

Proof of Claim 1. The right-hand side of this claim is given by $\tau(K_h, \{V(H)\}, V) = 1/(nm^{1/h})$.

For S of size $2 \leqslant s \leqslant (h-1)$, restricting the clique cover $\{V(H)\}$ to S gives a single s-clique, so

$$\begin{aligned} \tau(K_h, \{V(H)\}, S) &= 1/(nm^{1/s}) < 1/(nm^{1/h}), \text{ and} \\ \tau'(K_h, \{V(H)\}, S) &= 1/(nm^{1/s}) < 1/(nm^{1/h}). \end{aligned}$$

For S of size 1,

$$\begin{aligned} \tau(K_h, \{V(H)\}, S) &= 1/(nm) < 1/(nm^{1/h}), \text{ and} \\ \tau'(K_h, \{V(H)\}, S) &= 0 < 1(nm^{1/h}). \end{aligned} \qquad \square$$

Claim 2. *For the edge clique cover $E(H)$,*

$$\max_{S}\{\tau(K_h, E(H), S), \tau'(K_h, E(H), S)\} = \tau(K_h, E(H), V).$$

Proof of Claim 2. The right-hand side of the claim is given by $\tau(K_h, E(H), V) = 1/(n^{1/(h-1)}m^{1/2})$.

Let $S \subset V$, with $s = |S| < h$, and consider the restricted clique covers $E(H)[S]$ and $E(H)'[S]$. The cover $E(H)[S]$ is the edge cover of K_s plus a 1-clique for each of the $(h-s)$ external edges for each vertex of K_s, so

$$\tau(K_h, E(H), S) = 1/\left(n^{s/(s(s-1)+s(h-s))} m^{(s(s-1)/2+s(h-s))/(s(s-1)+s(h-s))}\right)$$

$$= 1/\left(n^{1/(h-1)} m^{(h-(s+1)/2)/(h-1)}\right)$$

$$\leqslant 1/\left(n^{1/(h-1)} m^{(h-(h/2))/(h-1)}\right), \text{ since } s \leqslant (h-1),$$

$$= 1/\left(n^{1/(h-1)} m^{h/(2(h-1))}\right)$$

$$< 1/(n^{1/(h-1)} m^{1/2}).$$

The restricted cover $E(H)'[S]$ is just the edge cover of K_s, and hence

$$\tau'(K_h, E(H), S) = 1/(n^{1/(s-1)} m^{1/2}) < 1/(n^{1/(h-1)} m^{1/2}). \qquad \square$$

Claim 3. *For any clique cover $\mathscr{C}$ of K_h in which each clique has size at least two, and for any fixed α,*

$$\tau(K_h, \mathscr{C}, V) \geqslant \tau(K_h, \{V(H)\}, V) \text{ or } \tau(K_h, \mathscr{C}, V) \geqslant \tau(K_h, E(H), V).$$

Proof of Claim 3. Let $\mathscr{C}$ be any clique cover of K_h in which each clique has size at least two, and denote the sizes of the cliques in $\mathscr{C} = \{C_1, C_2, \ldots, C_k\}$ by $r_1, r_2, \ldots, r_k$, respectively. The mechanism of this proof is a comparison of the linear function

$$\frac{h}{\Sigma\mathscr{C}} + \frac{|\mathscr{C}|}{\Sigma\mathscr{C}}\alpha,$$

for the exponent in $\tau(K_h, \mathscr{C}, V)$, to the functions for the exponents in $\tau(K_h, \{V(H)\}, V)$ and $\tau(K_h, E(H), V)$. These latter functions are given by

$$(*) \quad 1 + \tfrac{1}{h}\alpha \quad (\text{for } \{V(H)\}), \text{ and}$$

$$(**) \quad \tfrac{1}{h-1} + \tfrac{1}{2}\alpha \quad (\text{for } E(H)).$$

They are plotted in Figure 2, where $x(\alpha)$ represents the function $(h/\Sigma\mathscr{C}) + (|\mathscr{C}|/\Sigma\mathscr{C})\alpha$, with $\mathscr{C}$ being whichever clique cover is of interest.

The comparison will be made in two parts: first for $\alpha \in (0, \tfrac{2h}{h-1})$, and then for $\alpha \geqslant \tfrac{2h}{h-1}$. The border point $\alpha = \tfrac{2h}{h-1}$ is chosen because it is the value of α at which the functions $(*)$ and $(**)$ are equal. Here both function values are $1 + \tfrac{2}{h-1}$. For all smaller values of α, the function $(*)$ is the greater of the two, and for all larger α, $(**)$ is greater than $(*)$. Thus it is necessary only to compare $\tfrac{h}{\Sigma\mathscr{C}} + \tfrac{|\mathscr{C}|}{\Sigma\mathscr{C}}\alpha$ with $(*)$ for $\alpha \in (0, \tfrac{2h}{h-1})$, and with $(**)$ for $\alpha \geqslant \tfrac{2h}{h-1}$.

At $\alpha = 0$, $\tfrac{h}{\Sigma\mathscr{C}} + \tfrac{|\mathscr{C}|}{\Sigma\mathscr{C}}\alpha = \tfrac{h}{\Sigma\mathscr{C}}$, which is less than 1 since each vertex is in at least one of the cliques of $\mathscr{C}$. (In order to cover all edges, all end-point pairs must be included somewhere in the cover.)

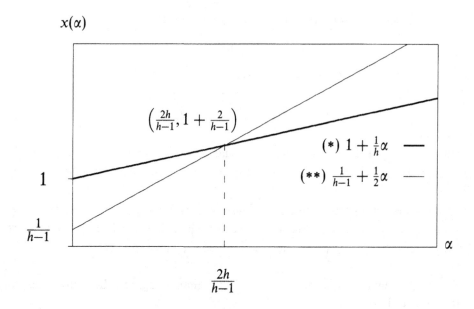

Figure 2 The exponent functions for $\tau(K_h, \{V(H)\}, V)$ and $\tau(K_h, E(H), V)$.

At $\alpha = \frac{2h}{h-1}$,

$$
\begin{aligned}
\frac{h}{\Sigma\mathscr{C}} + \frac{|\mathscr{C}|}{\Sigma\mathscr{C}}\alpha &= \frac{h}{\Sigma\mathscr{C}} + \frac{|\mathscr{C}|}{\Sigma\mathscr{C}}\left(\frac{2h}{h-1}\right) \\
&\leq \frac{h}{\Sigma\mathscr{C}} + \left(\frac{h+1}{2h} - \frac{h-1}{2\Sigma\mathscr{C}}\right)\left(\frac{2h}{h-1}\right) \quad \text{(see justification below)} \\
&= 1 + \frac{2}{h-1}.
\end{aligned}
$$

The substitution used to obtain the second inequality is justified as follows.

Let u be a vertex of $V(K_h)$ that appears in the fewest number of cliques of $\mathscr{C}$, and let t be the number of cliques in which u appears. Then

$$
\Sigma\mathscr{C} = \sum_{\{i:C_i \supset u\}} r_i + \sum_{\{i:C_i \not\supset u\}} r_i \geq [(h-1)+t] + [2(|\mathscr{C}|-t)], \tag{2.1}
$$

where $(h-1)$ counts all other vertices aside from u since they must each appear in some clique with u.

For any $v \in V(K_h)$,

$$
\begin{aligned}
\Sigma\mathscr{C} + |\{i : C_i \supset v\}| - (h-1) \\
&\geq \Sigma\mathscr{C} + t - (h-1) \quad \text{(by definition of } t \text{ as a minimum)} \\
&\geq [(h-1)+t] + [2(|\mathscr{C}|-t)] + t - (h-1) \quad \text{(by (2.1))} \\
&= 2|\mathscr{C}|.
\end{aligned}
$$

Taking the sum of both sides of this over all $v \in V(H)$ gives

$$h\Sigma\mathscr{C} + \Sigma\mathscr{C} - h(h-1) \geqslant 2h|\mathscr{C}|.$$

Divided by $2h\Sigma\mathscr{C}$ and rearranged, this yields

$$\frac{|\mathscr{C}|}{\Sigma\mathscr{C}} \leqslant \frac{h+1}{2h} - \frac{h-1}{2\Sigma\mathscr{C}},$$

as desired.

Now, since $\frac{h}{\Sigma\mathscr{C}} + \frac{|\mathscr{C}|}{\Sigma\mathscr{C}}\alpha$ is less than or equal to (*) at both $\alpha = 0$ and $\alpha = \frac{2h}{h-1}$, and the functions are linear, $\frac{h}{\Sigma\mathscr{C}} + \frac{|\mathscr{C}|}{\Sigma\mathscr{C}}\alpha$ is the minimum function throughout the interval $(0, \frac{2h}{h-1})$.

By the same analysis as above, $\frac{h}{\Sigma\mathscr{C}} + \frac{|\mathscr{C}|}{\Sigma\mathscr{C}}\alpha$ is less than or equal to (**) when $\alpha = \frac{2h}{h-1}$. Its slope is $\frac{|\mathscr{C}|}{\Sigma\mathscr{C}}$. By assumption, this is less than or equal to $1/2$, which is the slope of (**). Thus, for all $\alpha \geqslant \frac{2h}{h-1}$, $\frac{h}{\Sigma\mathscr{C}} + \frac{|\mathscr{C}|}{\Sigma\mathscr{C}}\alpha$ must be at most the value of (**). □

This is now enough information to apply the main subgraph theorems, since, for any $\mathscr{C}$ that is not $\{V(H)\}$ nor $E(H)$,

$$\max_S\{\tau(K_h, \mathscr{C}, S), \tau'(K_h, \mathscr{C}, S)\}$$

$$\geqslant \quad \tau(K_h, \mathscr{C}, V) \text{ (so assume } \mathscr{C} \text{ is irreducible)}$$

$$\geqslant \quad \max\{\tau(K_h, \{V(H)\}, V), \tau(K_h, E(H), V)\} \text{ (by Claim 3)}$$

$$= \quad \max_S\{\tau(K_h, \{V(H)\}, S), \tau'(K_h, \{V(H)\}, S), \tau(K_h, E(H), S), \tau'(K_h, E(H), S)\}$$

(by Claims 1 and 2).

Hence the minimum $\max_S\{\tau(K_h, \mathscr{C}, S), \tau'(K_h, \mathscr{C}, S)\}$ is always achieved by $\{V(H)\}$ or $E(H)$, giving

$$\min_{\mathscr{C}} \max_S \tau(K_h, \mathscr{C}, S) = \min\{\tau(K_h, \{V(H)\}, V), \tau(K_h, E(H), V)\}.$$

This implies that

$$\tau_{K_h,1} = \begin{cases} 1/(nm^{1/h}) & \text{for } \alpha \leqslant 2h/(h-1), \\ 1/(n^{1/(h-1)}m^{1/2}) & \text{for } \alpha \geqslant 2h/(h-1), \end{cases}$$

as desired. □

Corollary 7 (Complete bipartite graphs). *For the complete bipartite graph $K_{h,h}$, the appearance threshold (for all α) is*

$$\tau_{K_{h,h},1} = 1/(n^{1/h}m^{1/2}).$$

For the complete bipartite graph $K_{h,k}$ with $h > k$, the threshold is

$$\tau_{K_{h,k},1} = \begin{cases} 1/(n^{1/h}m) & \text{for } \alpha \leqslant \frac{h-k}{hk}, \\ 1/(n^{(h+k)/(2hk)}m^{1/2}) & \text{for } \alpha \geqslant \frac{h-k}{hk}. \end{cases}$$

Only the $h > k$ case will be proved here, since the result for $h = k$ follows similarly.

$x(\alpha)$

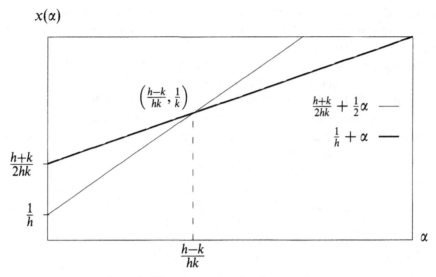

Figure 3   The exponent functions for τ_V and τ_{H_k}.

Since complete bipartite graphs are triangle-free, $E(H)$ is the only clique cover to be considered. The threshold function arises from considering this cover with vertex sets $V(H_k)$ (or any subset of $V(H_k)$), where H_k is the part of size k, and $V(H)$.

Let H_h refer to the vertex set of size h and H_k to the part of size k. It is straightforward to compute that

$$\tau_V = \tau(K_{h,k}, E(H), V(H)) \;=\; 1/\left(n^{\frac{h+k}{2hk}} m^{\frac{1}{2}}\right)$$

$$\tau_{H_h} := \tau(K_{h,k}, E(H), V(H_h)) \;=\; 1/\left(n^{\frac{1}{k}} m\right)$$

$$\tau_{H_k} := \tau(K_{h,k}, E(H), V(H_k)) \;=\; 1/\left(n^{\frac{1}{h}} m\right).$$

(Note that τ' need not be considered for these particular sets; for $V(H)$ it is the case that $\tau' = \tau$, and H_h and H_k have only 1-cliques so τ' is not applicable.)

The threshold function will be the maximum value of

$$\max\{\tau(K_{h,k}, E(H), S), \tau'(K_{h,k}, E(H), S)\}$$

over all vertex subsets of $V(H_{h,k})$. Since $h > k$, the above results imply that $\tau_{H_h} \leqslant \tau_{H_k}$ always, and τ_{H_h} can never be the maximum. Thus it can be ignored for the remainder of the proof.

Maximizing τ or τ' is the same as minimizing the exponent of n in the function denominator. For convenience, the exponent functions are referred to as $x(\alpha)$ and $x'(\alpha)$ for τ and τ' respectively. These exponents are also functions of the particular vertex set S being considered, and so may sometimes be written as, for instance, $x(\alpha, S)$ to draw attention to this. Figure 3 shows the $x(\alpha)$ functions for the cases $S = V(H)$ and $S = V(H_k)$ computed above.

As in the proof of the complete graph threshold, the linearity of the exponent functions is used. It will be shown that, at every possible value of α, $x(\alpha)$ and $x'(\alpha)$ for any S is at least as great as for $S = V(H)$ or $S = V(H_k)$. Thus the minimum is always $\min\{x(\alpha, V(H)), x(\alpha, V(H_k))\}$.

Consider any $S \subset V(H_k)$. Let $s = |S|$. Then $x(\alpha) = s/(sh) + 1 = (1/h) + 1$, so this is the same as for $S = V(H_k)$ and need not be considered.

Similarly, if $S \subset V(H_h)$ then $x(\alpha, S) = x(\alpha, H_h)$, so this case may be ignored.

Hence the only interesting case to be considered is that of a set S that has $s_h \geqslant 0$ vertices from part H_h and $s_h \geqslant 0$ vertices from part H_k. Clearly, $s_h \leqslant h$ and $s_k \leqslant k < h$. These facts are used in the proof.

For this S, the exponent functions are

$$x(\alpha) = \frac{s_h + s_k}{s_h k + s_k h} + \left(1 - \frac{s_k s_h}{s_h k + s_k h}\right)\alpha$$

$$x'(\alpha) = \frac{s_h + s_k}{2 s_h s_k} + \frac{1}{2}\alpha.$$

At $\alpha = 0$,

$$x = \frac{s_h + s_k}{s_h k + s_k h} \geqslant \frac{s_h + s_k}{s_h h + s_k h} = \frac{1}{h}, \quad \text{and}$$

$$x' = \frac{s_h + s_k}{2 s_h s_k} = \frac{1}{2 s_k} + \frac{1}{2 s_h} \geqslant \frac{1}{2h} + \frac{1}{2h} = \frac{1}{h},$$

so both functions are at least as great as the exponent for τ_{H_k}.

At $\alpha = \frac{h-k}{hk}$, which is the value of α at which the exponent functions of τ_V and τ_{H_k} intersect (see Figure 3),

$$x = \frac{s_h + s_k}{2 s_h s_k} + \left(1 - \frac{s_k s_h}{s_h k + s_k h}\right)\left(\frac{h-k}{hk}\right) = \frac{1}{k} + \frac{s_h(k - s_k)(h-k)}{(s_h k + s_k h)(hk)} \geqslant \frac{1}{k}, \quad \text{and}$$

$$x' = \frac{s_h + s_k}{s_h k + s_k h} + \frac{1}{2}\left(\frac{h-k}{hk}\right) = \frac{1}{2 s_k} + \frac{1}{2 s_h} + \frac{1}{2k} - \frac{1}{2h} \geqslant \frac{1}{2k} + \frac{1}{2h} + \frac{1}{2k} - \frac{1}{2h} = \frac{1}{k},$$

where $(1/k)$ is the value of $x(\alpha, V(H))$ and $x(\alpha, V(H_k))$ at $\alpha = (h-k)/(hk)$. Since all of the functions are linear, S must have $x(\alpha)$ and $x'(\alpha)$ at least as great as $x(\alpha, V(H_k))$ for the entire interval $(0, (h-k)/hk]$.

Moreover, $x(\alpha)$ and $x'(\alpha)$ are at least as great as the function $x(\alpha, V(H))$ at $\alpha = (h-k)/(hk)$. If it is possible to show that the slopes of $x(\alpha)$ and $x'(\alpha)$ are at least that of $x(\alpha, V(H))$, then this will imply that both functions have value at least $x(\alpha, V(H))$ for all $\alpha > (h-k)/(hk)$.

The slope condition turns out to be true, since $x'(\alpha)$ has slope $(1/2)$, which is equal to that of $x(\alpha, V(H))$, and $x(\alpha)$ has slope

$$\left(1 - \frac{s_k s_h}{s_h k + s_k h}\right) \geqslant \left(1 - \frac{s_k s_h}{s_h s_k + s_k s_h}\right) = \frac{1}{2}.$$

Thus

$$\tau_{K_{k,h},1} = \max_S\{\tau(K_{h,k}, E(H), S), \tau'(K_{h,k}, E(H), S)\} = \max\{\tau_V, \tau_{H_k}\}. \qquad \square$$

Corollary 8 (Triangle-free graphs). *If H is a triangle-free graph with h vertices and maximum degree Δ, then*

$$\tau_1(H) = \begin{cases} 1/(n^{1/\Delta}m^{1-(\varepsilon(Q)/|Q|\Delta)}) & \text{for } \alpha \text{ sufficiently small,} \\ 1/(n^{1/(2d^*(H))}m^{1/2}) & \text{for } \alpha \text{ sufficiently large,} \end{cases}$$

where Q is a subset of $V(H)$ that induces the most edges per vertex with the restriction that all of its vertices have degree Δ in H, and $d^*(H) = \max_{L \leqslant H} |E(L)|/|V(L)|$. $\qquad\square$

Note that finding the vertex set Q described above is equivalent to finding a vertex set achieving the maximum average degree for the graph induced on those vertices of H that are of degree Δ.

Corollary 9 (Trees). *If T is a tree with h vertices and maximum degree Δ, then*

$$\tau_1(T) = \begin{cases} 1/(n^{1/\Delta}m^{1-(r-1)/(r\Delta)}) & \text{for } \alpha \text{ sufficiently small,} \\ 1/(n^{h/(2h-2)}m^{1/2}) & \text{for } \alpha \text{ sufficiently large,} \end{cases}$$

where r is the size of a largest subtree induced on vertices all of which have degree Δ in T. $\qquad\square$

The proof of the small-α result is an application of the results proved for triangle-free graphs. It requires showing that a set Q of Δ-degree vertices of T that induces the most edges per vertex may be taken to be a set of Δ-degree vertices inducing a largest subtree in T, i.e., that

$$(\text{number of edges in } H[S])/|S| \leqslant (r-1)/r$$

holds for all S containing only vertices of degree Δ, where r is as defined above. The large-α threshold follows from Corollary 4, since $d^*(T) = \frac{h-1}{h}$.

Now consider the evolution of $G(n, m, p)$ as p is increased through a range of various functions for which $mp^2 \to 0$. In which order are the different subgraph thresholds reached? Are the cliques the first fixed graphs to arrive or the last? It turns out that the answer to this question depends on the value of α.

By examining the thresholds given above, one can see that when $\alpha = 3$ the cycle threshold is below that for all cliques on at least 4 vertices. However, when $\alpha = 1/2$ the clique threshold is below the threshold for cycles. This difference will be addressed more fully in a later paper.

3. Average case analysis of gate matrix circuit design

In this section we present an application of our theory to the design of integrated circuits via *gate matrix layouts* (GML).

A gate matrix layout is a stylized design for VLSI circuits [3, 4, 14]. The layout uses a grid-like representation, whose information can be encoded into a matrix of zeroes and ones that is much like the representation matrix introduced in Section 1.

The circuit components to be connected in a GML design are polysilicon lines serving as transistor gates and/or conductors ('gates'), and groups of transistor diffusions that

Table 2 Three ways to lay out the same GML circuit design. On the left is the original design. In the middle, we permuted columns 2 and 3. This allows us (right) to combine the first two rows.

1	2	3	4
a	—	a	
	b	—	b
c	—	c	c

1	2	3	4
a	a		
		b	b
c	c	—	c

1	2	3	4
a	a	b	b
c	c	—	c

associate with each other ('nets'). The GML grid is formed by laying all of the gate lines as vertical lines, and then placing horizontal tracks of transistor nets across them. There is a vertical gate line for each discrete input, and a transistor with that input must be placed over that gate. Additionally, all transistors associated in the same net must be placed in a common horizontal track. There may be more than one net in the same track, but only if the nets do not overlap (*i.e.*, their required inputs must be different, and must permit ordering such that one net is incident on all of the required gate lines before the next one in that track begins).

Although a layout that assigns a separate track for each net would be viable, the resulting circuit would undoubtedly be sparser than necessary. Since a chip with reduced surface area is cheaper to produce, we check for non-overlapping nets that can share tracks. We may do this for a given order of gate lines but, in addition, gate lines may be exchanged, carrying their transistors with them, and a different ordering of gates may allow for more nets to be placed in some tracks. See Table 2. Hence, the GML optimization problem is to find a permutation of the order of gates that minimizes the number of tracks required to lay out the circuit.

This problem may be stated equivalently in terms of matrices. We construct a gate matrix M by creating a column for each gate, and a row for each net. A '1' in position (i, j) represents a transistor that is in net i and must be placed on gate j. Roughly, we wish to find a column permutation of M so that the nets of 1s are packed densely in their individual rows (consecutive 1s are desirable).

Independent of the order of columns in M, we may use the matrix to construct the intersection graph G defined by

$$V = \{\text{set of nets}\},$$
$$E = \{(v_i, v_j) : v_i, v_j \text{ are incident on a common gate}\}.$$

The GML problem is known to be NP-hard [11] in general, but in the case when G is an *interval* graph[3] the problem is equivalent to that of finding a minimum colouring of G. Moreover, in the interval graph case the minimum colouring problem is easily solvable [8]. Thus, one might ask how likely it is that a gate matrix layout will have an interval graph as its associated graph G. We shall return to that question in the next section.

---

[3] Recall that G is called an *interval graph* provided we can assign to each vertex v of G a real interval I_v so that $vw \in E(G)$ exactly when $I_v \cap I_w \neq \emptyset$.

Consider a related decision problem known in the literature as the *k-GML problem.* Given an $m \times n$ gate matrix M and a fixed k, is there a permutation of columns of M such that the layout is possible in at most k tracks?

It has been recognized by Fellows and Langston (see [7]) that the the gate matrix layout with parameter k is equivalent to the path-width problem with parameter $k - 1$, that is, a graph represents a circuit with a k-track layout if and only if G has a path decomposition of width at most $k - 1$.

Although the original GML problem is in general very hard, Fellows and Langston [6] proved that the k-GML problem is, surprisingly, solvable in polynomial time (with respect to n). Their argument is based on the results of general theory of graph minors due to Robertson and Seymour (see [16]).

Suppose that matrix M represents the graph G defined as above. Notice that, if G corresponds to a matrix that satisfies the k-GML condition, then a corresponding matrix for any *minor*[4] of G satisfies k-GML also. This means that the family of graphs satisfying k-GML is closed under the minor ordering. Thus, according to Robertson–Seymour theory, the set of *obstructions* to k-GML, that is, minimal graphs under minor ordering that do not satisfy k-GML, must be finite. Since for every fixed graph H the problem of checking if H is a minor of G is solvable in polynomial time, k-GML can be decided in polynomial time as well. In fact, it can be decided in $O(n^2)$ (see [7]). The obstructions are known explicitly only when $k = 1, 2$ and 3. It is trivial to notice that for 1-GML an edge K_2 is such an obstruction. For 2-GML there are two such graphs: a triangle K_3 and subdivided star $K_{1,3}$ given as graph A in Figure 4 in the next subsection. For 3-GML, Kinnersley and Langston [12] were able to list all 110 obstruction graphs. They pointed out also that for $k = 4$ we may expect at least 122 million such graphs.

It is easy to notice, however, that K_k, the complete graph on k vertices, is an obstruction to $(k - 1)$-GML for $k \geqslant 2$. Hence, if the gate matrix graph G contains K_k as a subgraph, then we are able at least to give the following 'negative' type result.

For the average case analysis, Corollary 6 implies that in a random gate matrix layout we shall need, with probability tending to 1 as $n \to \infty$, at least k tracks if the probability $p \gg 1/(nm^{1/k})$ when $\alpha \leqslant 2k/(k - 1)$, or $p \gg 1/(n^{1/(k-1)}m^{1/2})$ for $\alpha \geqslant 2k/(k - 1)$.

What exactly is meant by an average case? In the GML problem statement, the appearance of 0s and 1s in the matrix M is not random, but fixed by the interconnection requirements of the circuit. However, a probabilistic model based on a probability p of having a 0 in any particular component of the matrix is considered in [3].

3.1. When is $G(n, m, p)$ likely to be an interval graph?

In the interest of finding the probability of easy cases for gate matrix layout analysis, we first apply our random intersection graph model and subgraph theorem to study the values of p for which $G(n, m, p)$ will with high probability be an interval graph. This is accomplished by studying the appearance and disappearance thresholds for the family of

---

[4] Recall that H is a *minor* of G if a graph isomorphic to H can be obtained from G by a series of two operations: taking a subgraph and contracting an arbitrary edge. Then we say that H is less than or equal to a graph G in the *minor order.*

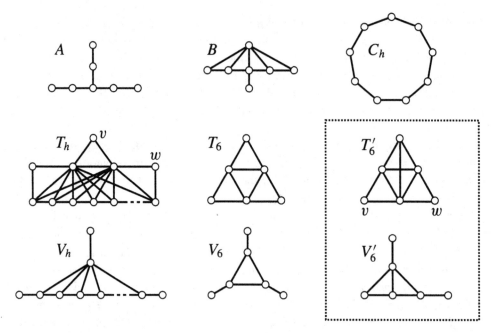

Figure 4   The graphs A, B, C_h ($h \geqslant 4$), T_h ($h \geqslant 6$), and V_h ($h \geqslant 6$) are the minimal non-interval graphs. The graph T_6' (resp. V_6') is an induced subgraph of every T_h ($h \geqslant 7$) (resp. V_h ($h \geqslant 7$)).

forbidden subgraphs that characterize the non-interval graphs. These forbidden subgraphs appear in Figure 4. The graphs A, B, C_h ($h \geqslant 4$), T_h ($h \geqslant 6$), and V_h ($h \geqslant 6$) are not interval graphs, but all of their proper induced subgraphs are interval graphs. Indeed, these are the only such graphs (see [13]; see also [8]). Since induced subgraphs of interval graphs are interval, it follows that G is an interval graph if and only if it induces none of the graphs shown in the main part of Figure 4 as subgraphs.

We are thus interested in the range of p that is below all appearance thresholds for forbidden subgraphs, and the range that is above all disappearance thresholds for these subgraphs. The forbidden subgraph family can be partitioned into two individual graphs, A and B, and three infinite families of subgraphs. For each fixed value of α, the appearance threshold for forbidden subgraphs is the minimum of the appearance thresholds for all of the individual forbidden subgraphs. We begin by showing that we need only consider the smallest member of each infinite family, since its appearance threshold is the minimum threshold for its family.

The individual graph thresholds given here result from applications of Theorem 3, and were obtained with the aid of a computer program (to find the maximum and minimum threshold function values depending on α).

As given in Corollary 5, the appearance threshold for a cycle on h vertices, where h is greater than or equal to 4, is

$$\tau_1(C_h) = 1 / \left(n^{\frac{1}{2}} m^{\frac{1}{2}} \right).$$

So we need only consider C_4 for this family, with threshold as above.

For the family of graphs T_h, $h \geqslant 6$, we compare all larger members to T_6. Consider the graph T_6' in Figure 4. It is an induced subgraph of each member of the T_h-family for which $h \geqslant 7$ (vertices v and w correspond as indicated). One checks that the appearance threshold for T_6' is greater than or equal to that of T_6, for all α (see [19]). Thus, it suffices (for the T-family) to consider only T_6, whose threshold is

$$
\tau_1(T_6) = \begin{cases} 1/\left(n^{\frac{1}{2}}m^{\frac{1}{2}}\right) & \text{for } \alpha \leqslant 1, \\ 1/\left(n^{\frac{2}{3}}m^{\frac{1}{3}}\right) & \text{for } 1 \leqslant \alpha \leqslant 2, \\ 1/\left(n^{\frac{1}{3}}m^{\frac{1}{2}}\right) & \text{for } \alpha \geqslant 2. \end{cases}
$$

Now consider the graph V_6' in Figure 4, noting that it is an induced subgraph of V_h for all $h \geqslant 7$. Again, one checks that the appearance threshold for V_6' is always at least that of V_6, so the appearance thresholds of V_h, $h \geqslant 7$ are greater than or equal to that of V_6. It thus suffices (for the V-family) to consider only V_6, whose threshold is

$$
\tau_1(V_6) = \begin{cases} 1/\left(n^{\frac{1}{2}}m^{\frac{2}{3}}\right) & \text{for } \alpha \leqslant \frac{3}{4}, \\ 1/\left(n^{\frac{2}{3}}m^{\frac{4}{9}}\right) & \text{for } \frac{3}{4} \leqslant \alpha \leqslant 3, \\ 1/\left(n^{\frac{1}{2}}m^{\frac{1}{2}}\right) & \text{for } \alpha \geqslant 3. \end{cases}
$$

For the two individual forbidden subgraphs, A and B, we have

$$
\tau_1(A) = \begin{cases} 1/\left(n^{\frac{1}{3}}m\right) & \text{for } \alpha \leqslant \frac{1}{3}, \\ 1/\left(n^{\frac{4}{9}}m^{\frac{2}{3}}\right) & \text{for } \frac{1}{3} \leqslant \alpha \leqslant \frac{5}{6}, \\ 1/\left(n^{\frac{7}{12}}m^{\frac{1}{2}}\right) & \text{for } \alpha \geqslant \frac{5}{6}, \end{cases}
$$

and

$$
\tau_1(B) = \begin{cases} 1/\left(n^{\frac{1}{3}}m^{\frac{5}{6}}\right) & \text{for } \alpha \leqslant \frac{1}{5}, \\ 1/\left(n^{\frac{3}{8}}m^{\frac{5}{8}}\right) & \text{for } \frac{1}{5} \leqslant \alpha \leqslant \frac{1}{3}, \\ 1/\left(n^{\frac{5}{12}}m^{\frac{1}{2}}\right) & \text{for } \frac{1}{3} \leqslant \alpha \leqslant \frac{5}{6}, \\ 1/\left(n^{\frac{5}{11}}m^{\frac{5}{11}}\right) & \text{for } \frac{5}{6} \leqslant \alpha \leqslant \frac{6}{5}, \\ 1/\left(n^{\frac{1}{2}}m^{\frac{5}{12}}\right) & \text{for } \frac{6}{5} \leqslant \alpha \leqslant 2, \\ 1/\left(n^{\frac{1}{3}}m^{\frac{1}{2}}\right) & \text{for } \alpha \geqslant 2. \end{cases}
$$

Taking the minimum of all of these thresholds, we find that the appearance threshold for the forbidden subgraphs is

$$\tau_1 \text{ (forbidden subgraph set)} = \begin{cases} 1/\left(n^{\frac{1}{2}}m^{\frac{2}{3}}\right) & \text{for } \alpha \leqslant \frac{3}{4}, \\ 1/\left(n^{\frac{2}{3}}m^{\frac{4}{9}}\right) & \text{for } \frac{3}{4} \leqslant \alpha \leqslant \frac{3}{2}, \\ 1/\left(n^{\frac{7}{12}}m^{\frac{1}{2}}\right) & \text{for } \alpha \geqslant \frac{3}{2}. \end{cases}$$

This means that for $p = \tau/\omega$, with τ as above, $G(n,m,p)$ is with high probability an interval graph. If $p = \omega\tau$ (and $mp^2 \nrightarrow \infty$), then G is with high probability not an interval graph.

When $mp^2 \rightarrow \infty$ we approach the end of the evolution and we consider the threshold when none of the forbidden subgraphs appear. This problem is much simpler. Consider the forbidden subgraph C_4. By Theorem 3 we have $\tau_2(C_4) = \sqrt{2\log n/m}$, which is the same as the threshold for $G(n,m,p)$ becoming a complete graph! Thus, in order to have probability tending to 1 that $G(n,m,p)$ an interval graph (for 'large' p), we need to choose p large enough that $G(n,m,p)$ is almost surely complete, that is, $p = \sqrt{\frac{2\log n + \omega}{m}}$.

References

[1] Bollobás, B. (1985) *Random Graphs*, Academic.

[2] Das, S. Expected number of tracks for gate matrix, Dept. of Computer Science, University of North Texas.

[3] Das, S., Deo, N., and Prasad, S. (1989) *Gate Matrix Layout Revisited: Algorithmic Performance and Probabilistic Analysis*, Vol. 405 of *Lecture Notes in Computer Science*, Springer, pp. 280–290.

[4] Deo, N., Krishnamoorthy, M., and Langston, M. (1987) Exact and approximate solutions for the gate matrix layout problem, *IEEE Trans. Computer-Aided Design* **CAD-6** 79–84.

[5] Erdős, P. and Rényi, A. (1960) On the evolution of random graphs. *MTA Mat. Kut. Int. Kozl.* **5** 17–61.

[6] Fellows, M. and Langston, M. (1987/88) Nonconstructive advances in polynomial-time complexity. *Information Processing Letters* **26** 157–162.

[7] Fellows, M. and Langston, M. (1994) On search, decision, and the efficiency of polynomial-time algorithms. *J. Computer and System Sciences* **49** 769–779.

[8] Golumbic, M. (1980) *Algorithmic Graph Theory and Perfect Graphs*, Academic.

[9] Karoński, M. (1995) Random graphs. In *Handbook of Combinatorics* (R. L. Graham, M. Grötschel and L. Lovász, eds), Vol. 1, Elsevier, pp. 351–380.

[10] Karoński, M. and Ruciński, A. (1989) Small subgraphs of *k*-partite random graphs. In *Combinatorial Mathematics: Proc. Third International Conference, New York, 1985*, 555, New York Academy of Sciences, pp. 230–240.

[11] Kashiwabara, T. and Fujisawa, T. (1979) An NP-complete problem on interval graphs. *Proc. IEEE Symposium on Circuits and Systems*, 82–83.

[12] Kinnersley, N. and Langston, M. (1994) Obstruction set isolation for the gate matrix layout problem. *Discrete Appl. Math.* **54** 169–213.

[13] Lekkerkerker, C. B. and Boland, J. C. (1962) Representation of a finite graph by a set of intervals on the real line. *Fund. Math.* **51** 45–64.

[14] Lopez, A., and Law, H.-F. (1980) A dense gate matrix layout method for MOS VLSI. *IEEE Trans. Electron Devices* **ED-27** 1671–1675.

[15] Marczewski, E. (1945) Sur deux propriétés des classes d'ensembles. *Fund. Math.* **33** 303–307.

[16] Robertson, N. and Seymour, P. D. (1985) Graph minors – A Survey. In *Surveys in Combinatorics* (I. Anderson, ed.), Cambridge University Press, Cambridge, pp. 153–171.

[17] Scheinerman, E. A. (1988) Random interval graphs. *Combinatorica* **8** 357–371.

[18] Scheinerman, E. A. (1990) An evolution of interval graphs. *Discrete Math.* **82** 287–302.

[19] Singer, K. (1995) *Random Intersection Graphs*, PhD thesis, Johns Hopkins University.

Combinatorics, Probability and Computing (1999) 8, 161–176.
© 1999 Cambridge University Press

The Blow-up Lemma

JÁNOS KOMLÓS

Department of Mathematics, Rutgers University, New Brunswick, NJ 08903, USA
(e-mail: komlos@math.rutgers.edu)

To the memory of Paul Erdős

Extremal graph theory has a great number of conjectures concerning the embedding of large sparse graphs into dense graphs. Szemerédi's Regularity Lemma is a valuable tool in finding embeddings of small graphs. The Blow-up Lemma, proved recently by Komlós, Sárközy and Szemerédi, can be applied to obtain approximate versions of many of the embedding conjectures. In this paper we review recent developments in the area.

This paper is based on my lectures at the DIMANET Mátraháza Workshop, October 22–28, 1995. On my transparencies, I wrote, 'For more details see the survey of Komlós–Simonovits in *Paul Erdős is 80* [20]. Solutions to the conjectures mentioned today will be presented in the Bolyai volume *Paul Erdős is 90*.' As you can tell, at that time I expected EP (who was sitting in the front row) to live to be 90 and more. The loss is obvious to all of us, and it will certainly deepen further in time.

1. Introduction

Our concern in this paper is how Szemerédi's Regularity Lemma can be applied to packing (or embedding) problems. In particular, we discuss a lemma that is a powerful weapon in proving the existence of embeddings of large sparse graphs into dense graphs.

After a brief passage in which we fix the notation, we start in Section 2 by recalling some of the fundamental results and conjectures. Section 3 is about the Regularity Lemma itself; we also demonstrate its power by reconstructing the elegant proof of Ruzsa and Szemerédi for Roth's theorem on arithmetic progressions of length 3.

In Section 4 we show how the Regularity Lemma can be applied to embedding a large sparse graph into a dense graph. We describe a method of constructing such an embedding in five separate phases; in this way we isolate the application of the Regularity Lemma from the other issues that typically need to be dealt with.

Section 5 is about the Blow-up Lemma, a result designed to overcome some of the unpleasant technical difficulties often arising in applications of Szemerédi's Regularity Lemma. The main benefit of the Blow-up Lemma is that in suitable contexts we may regard sufficiently regular pairs of subsets of vertices as spanning a complete bipartite subgraph. We will illustrate this principle by a concrete example and mention recent progress concerning major problems in the area.

Definitions and notation

We will write $v(G)$ for the order of G, G_n for an n-graph (order n), and $e(G)$ for the size of G, that is, for the number of edges in G; we denote the minimum, the maximum, and the average degree in G by $\delta(G), \Delta(G)$, and $t(G)$, respectively. Also, $\deg(v)$ is the degree of v, the number of edges incident with v, and $\deg(v, X)$ is the number of edges from v to X. We write $\chi(G)$ for the chromatic number of G, and $\overline{G}$ for the complement of G. Given a set U of vertices, $G|_U$ is the restriction of G to U. As usual, $H \subset G$ means that H is a subgraph of G, or at least that G has a subgraph isomorphic to H; we also say that H embeds into G, that is, there is a one-to-one map (injection) $\varphi : V(H) \to V(G)$ such that $\{x, y\} \in E(H)$ implies $\{\varphi(x), \varphi(y)\} \in E(G)$. We denote by $\|H \to G\|$ the number of (labelled, but not necessarily induced) isomorphic copies of H in G. The graphs $G_1 = (V, E_1)$ and $G_2 = (V, E_2)$ *pack* (or *can be packed*) if they can be placed without overlapping edges. More precisely, they pack if there is a bijection $\varphi : V \to V$ such that $\{x, y\} \in E_1$ implies $\{\varphi(x), \varphi(y)\} \notin E_2$. This is clearly the same as $G_1 \subset \overline{G_2}$.

Finally, given a graph R and a positive integer t, let us write $R(t)$ for the graph obtained by replacing the vertices $v_i \in V(R)$ by (pairwise disjoint) t-sets V_i and replacing the edges $\{v_i, v_j\} \in E(R)$ by complete bipartite graphs between V_i and V_j.

2. Classical embedding results and conjectures

The overall form of all such results is the same: if G_1 and G_2 are 'small' then they pack. But I will mostly use the embedding language: if H is 'small' and G is 'large' then H can be embedded into G. Much of classical extremal graph theory can be stated in these loose terms and it is the various specific meanings of 'small' and 'large' in different theorems that give them different flavours.

Below I give a short list of some classical embedding results and conjectures. This list is not at all complete and I usually mention only the simplest forms of these results. (For relevant definitions and more information see Bollobás's book [7] and Simonovits's survey [25].) The list is divided into five categories according to the order of H (the graph to be embedded) relative to n, the order of the *host-graph* G:

(1) **fixed** H
(2) H **of order** $o(n)$
(3) **medium-size** H (*i.e.*, $v(H) \approx cn$, $0 < c < 1$)
(4) **large** H (order $(1 - o(1))n$ – almost perfect packing)
(5) **full-size** H (order n – spanning subgraphs).

2.1. Fixed H

The first theorem here, by Turán in 1941 [28], was the starting point of extremal graph theory, and the second one, by Erdős and Stone in 1946 [14] and Erdős and Simonovits in 1966 [12], summarizes its most important feature.

Theorem 1 (Turán [28]). *If $e(G_n) > \left(1 - \frac{1}{r-1}\right) \frac{n^2}{2}$ then $K_r \subset G_n$.*

Theorem 2 (Erdős, Stone and Simonovits [14, 12]). *For every $\varepsilon > 0$ and H there is an n_0 such that, if $n \geqslant n_0$ and*

$$e(G_n) > \left(1 - \frac{1}{\chi(H) - 1} + \varepsilon\right) \frac{n^2}{2},$$

then $H \subset G_n$. In fact, $\|H \to G_n\| > \alpha n^{v(H)}$, where $\alpha = \alpha(H, \varepsilon) > 0$.

Hence, for the Turán number $ex(n, H)$, one has

$$\lim_{n \to \infty} \frac{ex(n, H)}{\binom{n}{2}} = 1 - \frac{1}{\chi(H) - 1}.$$

The main message from Theorem 2 is that in extremal graph theory $\chi(H)$ is the relevant quantity. Contrast this to random graph theory, where the critical quantity is the average degree $t(H)$ or, more precisely, the maximum average degree $\mathrm{MAD}(H) = \max_{H' \subseteq H} t(H')$. For a fixed H, one needs $e(G_n) \sim c(H) n^{2 - 2/\mathrm{MAD}(H)}$ to ensure that most n-graphs G_n contain a subgraph isomorphic to H.

2.2. H of order $o(n)$

The following two theorems show the tremendous power of the Regularity Lemma in extremal graph theory. They estimate the Turán and Ramsey numbers for very general classes of graphs. They are both from the same landmark paper by Chvátal, Rödl, Szemerédi and Trotter in 1983 [8] (the first one is only implicit in the paper and is quoted by Alon, Duke, Leffman, Rödl and Yuster [3] as such).

Theorem 3 (Chvátal, Rödl, Szemerédi and Trotter [8]). *For every Δ and $c > 0$ there is an $\alpha > 0$ such that, if $e(G_n) > cn^2$ and H is any bipartite graph of order less than αn and maximum degree $\Delta(H) \leqslant \Delta$, then $H \subset G_n$.*

Theorem 4 (Ramsey numbers of bounded degree graphs are linear [8]). *For every Δ there is an $\alpha > 0$ such that, if G is an n-graph and H is any graph of order less than αn and maximum degree $\Delta(H) \leqslant \Delta$, then $H \subset G$ or $H \subset \overline{G}$. In other words, $r(H) \leqslant c(\Delta(H))v(H)$ for all graphs H.*

2.3. Medium-size H

The classical conjectures in this category are about embedding trees. In this subsection, T denotes trees while G are arbitrary graphs. The greedy algorithm proves the following trivial claim: $\delta(G) \geqslant e(T)$ implies $T \subset G$. A classical conjecture of Erdős and T. Sós from 1963 [13] says that minimum degree here can be replaced by average degree.

Conjecture 5 (Erdős and Sós [13]).   *If $t(G) > e(T) - 1$ then $T \subset G$.*

That is, if k-stars are forced, then so are all trees of size k.

(Since the above trivial claim implies that $T \subset G$ whenever $t(G) \geqslant 2e(T)$, the conjecture amounts to gaining a factor 2 in the condition for $t(G)$.)

The following conjecture, on the other hand, is fairly recent.

Conjecture 6 (Loebl [11]).   *If G_n has at least $n/2$ vertices of degree at least $n/2$, then it contains all trees of size at most $n/2$.*

And here is a generalization by Komlós and T. Sós (see [2]). It says that the average degree in Conjecture 5 can probably be replaced by the median of the degrees (median degree, for short). (The number x is a median of the numbers $d_1, \ldots, d_n$ if at least half of them exceed x and at least half of them do not exceed x.)

Conjecture 7 (Komlós and Sós [2]).   *A graph G contains all trees of size not exceeding the median degree of G.*

I am not going to discuss these tree conjectures in the rest of the paper.

2.4. Large H
The interpretation of 'large' G is different here. Large $e(G_n)$ is no longer sufficient: we need lower bounds on all degrees.

Theorem 8 (Alon and Yuster [5]).   *For every $\varepsilon > 0$ and H there is an n_0 such that, if $n \geqslant n_0$ and*

$$\delta(G_n) \geqslant \left(1 - \frac{1}{\chi(H)}\right) n,$$

then G_n contains at least $(1 - \varepsilon)n/|V(H)|$ vertex-disjoint copies of H.

(Note the change of $\chi - 1$ to χ. We want a very large H-factor; for a number between the two we would still get an H-factor covering a certain positive proportion of the vertices of G_n.)

The following two easy theorems are about embedding almost spanning trees, cycles and paths. For stronger statements see the next subsection.

Theorem 9.   *For every $\varepsilon > 0$ there are $\alpha > 0$ and n_0 such that, if $n \geqslant n_0$, $\delta(G_n) \geqslant \frac{1}{2} n$, and T is a tree with $e(T) < (1 - \varepsilon)n$ and $\Delta(T) < \alpha n$, then $T \subset G_n$.*

Theorem 10.   *For every $\varepsilon > 0$ there is an n_0 such that, if $n \geqslant n_0$, $\delta(G_n) \geqslant \frac{2}{3} n$, $|V(H)| < (1 - \varepsilon)n$, and $\Delta(H) \leqslant 2$, then $H \subset G_n$.*

2.5. Full-size H
This is certainly the most interesting and hardest class.

Table 1

Powers of a cycle	Bounded degree trees	Bounded degree graphs
If $\delta(G_n) \geqslant n/2$ then G_n contains a Hamiltonian cycle	If $\delta(G_n) \geqslant (n-1)/2$ then G_n contains a Hamiltonian path	If $\delta(G_n) \geqslant (n-1)/2$ then G_n contains a 'perfect' matching
Pósa & Seymour (Conjecture 11)	Bollobás (Conjecture 12)	Hajnal & Szemerédi (Theorem 13) Bollobás & Eldridge (Conjecture 14) Alon & Yuster (Conjecture 15)

The mother of all full-size packing problems is Dirac's theorem of 1952 [10]. Here are various forms and problems that can be considered their generalizations. In Table 1 we arrange them in tabular form: the column headers indicate the main feature of the graph H to be embedded. Below, we state them one by one.

Conjecture 11 (Pósa and Seymour [24]). *If $\delta(G_n) \geqslant \frac{r}{r+1}n$, then G_n contains the rth power of a Hamiltonian cycle. (This would actually imply Theorem 13 below.)*

Conjecture 12 (Bollobás [7]). *For every $\varepsilon > 0$ and Δ there is an n_0 such that, if $n \geqslant n_0$, $\delta(G_n) > (\frac{1}{2} + \varepsilon)n$, and T is a tree of order n with maximum degree $\Delta(T) \leqslant \Delta$, then $T \subset G_n$.*

The following beautiful (and hard!) classical theorem is one of the main tools in embedding algorithms. (The original formulation was about colouring. The case $r = 3$ was solved by Corrádi and Hajnal [9].)

Theorem 13 (Hajnal and Szemerédi [15]). *If $\delta(G_n) \geqslant (1 - \frac{1}{r})n$, then G_n contains $\lfloor n/r \rfloor$ vertex-disjoint copies of K_r.*

Theorem 13 would follow from the following very hard packing conjecture.

Conjecture 14 (Bollobás and Eldridge [7]). *If $v(G_1) = v(G_2) = n$ and*

$$(\Delta(G_1) + 1)(\Delta(G_2) + 1) \leqslant n + 1,$$

then G_1 and G_2 pack.

(For $\Delta(G_1) = 2$, this is about the union of cycles and paths, and was conjectured by Sauer and Spencer [23], and proved by Aigner and Brandt [1] as well as by Alon and Fischer [4].)

Theorems 2 and 8 suggest that the chromatic number should be the right quantity, not the maximum degree. Here is another beautiful conjecture expressing this.

Conjecture 15 (Alon and Yuster). *For every H there are n_0 and K such that, if $n \geqslant n_0$ and*

$$\delta(G_n) \geqslant \left(1 - \frac{1}{\chi(H)}\right) n, \tag{2.1}$$

then G_n contains at least $(n - K)/|V(H)|$ vertex-disjoint copies of H.

Alon and Yuster [6] give an approximate solution by showing that, for every $\varepsilon > 0$ and H, there is an n_0 such that, if $n \geqslant n_0$,

$$\delta(G_n) \geqslant \left(1 - \frac{1}{\chi(H)} + \varepsilon\right) n, \quad \text{and} \quad |V(H)| \text{ divides } n,$$

then G_n can be perfectly tiled by copies of H.

Remark. The conjecture is false for $\varepsilon = 0$, that is, without the extra term εn. However, Alon and Yuster conjectured that an extra $C = C(H)$ suffices. In 1990, Erdős and Faudree conjectured that this latter conjecture is true with $C(H) = 0$ for the special case $H = C_4$.

Since chromatic number is the relevant quantity for embedding problems, the following would be a natural extension of Theorem 8.

Replace $\Delta(G_1) + 1$ in Conjecture 14 by $\chi(G_1)$. That is, for every $\varepsilon > 0$, r and Δ there is an n_0 such that, if $v(G_1) = v(G_2) = n \geqslant n_0$, $\chi(G_1) \leqslant r$, $\Delta(G_1) \leqslant \Delta$, and $\chi(G_1)\Delta(G_2) \leqslant (1 - \varepsilon)n$, then G_1 and G_2 pack. Alternatively, in embedding form: if $v(H) = v(G) = n \geqslant n_0$, $\chi(H) \leqslant r$, $\Delta(H) \leqslant \Delta$, and $\delta(G) \geqslant \left(1 - \frac{1}{r} + \varepsilon\right) n$, then $H \subset G$.

Unfortunately, this is false even for $r = 2$. Let G be the union of two cliques with only a little overlap, and H a random bipartite graph (an expander).

Why did it work for copies of H? It may have something to do with band-width. Let us write $w(G)$ for the band-width of the graph G.

Conjecture 16 (Bollobás and Komlós). *For every $\varepsilon > 0$, r and Δ, there are $\alpha > 0$ and n_0 such that, if $v(H) = v(G) = n \geqslant n_0$, $\chi(H) \leqslant r$, $\Delta(H) \leqslant \Delta$, $w(H) < \alpha n$, and $\delta(G) \geqslant \left(1 - \frac{1}{r} + \varepsilon\right) n$, then $H \subset G$.*

Another way to sharpen Theorem 8 and Conjecture 15 is to observe that the minimum degree $(1 - 1/r)n$ (where $r = \chi(H)$) is necessary only for graphs similar to K_r. For example, a special case of a famous conjecture of El-Zahar [29] says that when $H = C_k$ (cycle of length k) the minimum degree needed to get an H-factor in an n-graph is $n/2$ when k is even (as it should be) but only $(1/2 + 1/(2k))n$ when k is odd, rather than $(2/3)n$ as in Conjecture 15. The question naturally arises as to what quantity χ' can replace χ in (2.1).

An obvious obstruction is that one cannot embed a graph H into a graph G_n of lower chromatic number. That is, we certainly cannot replace $r = \chi$ in (2.1) with anything less than $\chi - 1$, as the example of $G_n = K_{r-1}(n/(r-1))$ shows, but there may be room for improvement between χ and $\chi - 1$.

Here is a more subtle obstruction. For an r-chromatic graph H, let us define the i-independence number α_i as the maximum possible sum of i colour-class sizes in any r-colouring of H, and the i-chromatic number as

$$\chi_i(H) = i\,\frac{v(H)}{\alpha_i(H)}.$$

(Thus, $\chi_i > i$ for $i < \chi$ and $\chi_i = i$ for $i \geqslant \chi$.)

Now, let G_n be a complete r-partite graph with the following colour-class sizes (n is large): i classes of size greater than n/χ_i each, and one left-over class (of size less than $n(1 - i/\chi_i)$). Then, clearly, $H \not\subset G_n$. Thus, one could attempt to replace χ in (2.1) by the largest of the numbers χ_i, $i < \chi$. But it is easy to see that

$$\frac{v(H)}{\alpha(H)} \leqslant \chi_1 \leqslant \chi_2 \leqslant \ldots \leqslant \chi_r = \chi(H),$$

and thus χ_{r-1} is the largest one. Hence I suggest the following quantity:

$$\chi'(H) = (\chi(H) - 1)v(H)/\alpha'(H), \tag{2.2}$$

where α' is the maximum possible sum of $\chi(H) - 1$ colour-class sizes in any colouring of H with $\chi(H)$ colours. (It is easy to see that $\chi - 1 < \chi' \leqslant \chi$, and $\chi' = \chi = r$ can hold only for graphs in which every r-colouring has equal colour-sizes.) In other words, I propose the following conjecture.

Conjecture 17. *For every H there is a K such that, if*

$$\delta(G_n) \geqslant \left(1 - \frac{1}{\chi'(H)}\right) n, \tag{2.3}$$

then G_n contains an H-factor covering all but at most K vertices.

The above example shows that the conjecture – if true – is best possible for any H.

All the above were about two types of H: *bounded degree graphs* (fixed H, or the union of its copies, powers of a Hamilton cycle, *etc.*) and *trees* (forests) (with degrees up to αn). In the rest of the paper I'll concentrate on bounded degree subgraphs.

3. The Regularity Lemma

In a bipartite graph $G = (A, B, E)$ (A and B are the colour classes), the *density* is defined as

$$d(A, B) = \frac{e(A, B)}{|A| \cdot |B|}.$$

We say that $G = (A, B, E)$ is an ε-*regular* pair (or more often we say that (A, B) is an ε-regular pair) if

$$X \subset A, |X| > \varepsilon|A|, \; Y \subset B, |Y| > \varepsilon|B| \quad \text{imply} \quad |d(X, Y) - d(A, B)| < \varepsilon.$$

We say that $G = (A, B, E)$ is an (ε, δ)-*super-regular* pair if

$$X \subset A, |X| > \varepsilon|A|, \; Y \subset B, |Y| > \varepsilon|B| \quad \text{imply} \quad e(X, Y) > \delta|X||Y|;$$

furthermore, $\deg(a) > \delta|B|$ for all $a \in A$, and $\deg(b) > \delta|A|$ for all $b \in B$.

The following fact is the most important property of regular pairs.

Fact. (Most degrees into a large set are large) Let (A, B) be an ε-regular pair with density d. Then (with $\delta = d - \varepsilon$), for any $Y \subset B$, $|Y| > \varepsilon|B|$,

$$\#\{x \in A : \deg(x, Y) \leqslant \delta|Y|\} \leqslant \varepsilon|A|.$$

Naturally, a similar bound holds for the number of vertices x with $\deg(x, Y) \geqslant (d+\varepsilon)|Y|$. In particular, most pairs of vertices in A have about the right number of common neighbours (about $d^2|B|$) and the same holds for B. This latter property turns out to be equivalent to regularity ([26] on pseudo-random graphs). That is, *the property that a bipartite graph is a regular pair is equivalent to its bipartite adjacency matrix being close to the sum of an orthogonal matrix and a constant matrix.*

An ε-regular pair of density d nicely imitates random bipartite graphs, provided ε is small enough in terms of d (often $\varepsilon \leqslant d/2$ is enough). Let me list a few such random-like properties in an informal way: all small trees are subgraphs; all small bipartite graphs of bounded degree are subgraphs; large chunks of a regular pair are regular; an (ε, δ)-super-regular pair (with ε small enough) has diameter at most 4.

Now, Szemerédi's Regularity Lemma (Theorem 18 below) says that *every (dense) graph is the union of a small number of regular pairs plus a little noise.*

Theorem 18 (Regularity Lemma [27]). *For every $\varepsilon > 0$ and m there are M and n_0 with the following property: for every graph $G = (V, E)$ with $n \geqslant n_0$ vertices there is a partition of the vertex set into $k + 1$ classes (clusters)*

$$V = V_0 + V_1 + V_2 + \ldots + V_k,$$

such that

- $m \leqslant k \leqslant M$
- $|V_1| = |V_2| = \ldots = |V_k|$
- $|V_0| < \varepsilon n$
- *all but at most εk^2 of the pairs $\{V_i, V_j\}$ are ε-regular.*

This is not a very transparent theorem, but it grows on you with time. The following is a simple consequence.

Theorem 19 (Regularity Lemma: degree form). *For every $\varepsilon > 0$ there is an $M = M(\varepsilon)$ such that, if $G = (V, E)$ is any graph and $d \in [0, 1]$ is any real number, then there is a partition of the vertex set V into $k + 1$ clusters $V_0, V_1, \ldots, V_k$, and there is a subgraph $G' \subset G$ with the following properties:*

- $k \leqslant M$
- $|V_0| \leqslant \varepsilon|V|$
- *all clusters V_i, $i \geqslant 1$, are of the same size $N \leqslant \lceil \varepsilon|V| \rceil$*
- $\deg_{G'}(v) > \deg_G(v) - (d + \varepsilon)|V|$ *for all $v \in V$*
- *for each $i \geqslant 1$, $G'|_{V_i}$ is empty*
- *all pairs $G'|_{V_i \times V_j}$ ($1 \leqslant i < j \leqslant k$) are ε-regular, each with a density either 0 or exceeding d.*

A graph, such as $G'' = G'|_{V-V_0}$, satisfying the last two properties will be called *pure*.

The reduced graph

Given an ε-regular partition and a number $d \geqslant 0$, we define the *reduced graph* R on $\{V_1, V_2, \ldots, V_k\}$ by connecting V_i and V_j if (V_i, V_j) is an ε-regular pair with density $> d$.

The graph R reflects many properties of G. For instance, R contains a triangle if and only if G contains many of them (cn^3). For much stronger properties see the Key Lemma and the Blow-up Lemma later.

For the graph G'' in the *degree form*, V_i and V_j are adjacent in the reduced graph if and only if there are *any* edges of G'' between them. But then, automatically, there are many $(dN^2 = cn^2)$ edges of G'' between them, since the density $d(V_i, V_j) > d$. Here is a beautiful illustration of this.

Theorem 20 (Roth [21]). *Every subset of $\mathbb{Z}$ of positive density contains an arithmetic progression of three terms.*

Roth's original proof uses Fourier analysis. Here is an elegant proof by Ruzsa and Szemerédi from 1976 (see [22]), based on the Regularity Lemma.

Proof. Let $R = r_3(n)$, and let $(1 \leqslant)a_1 < a_2 < \cdots < a_R(\leqslant n)$ be a maximum length sequence without a three-term arithmetic progression. Define a bipartite graph $G = G_{5n} = (A, B, E)$ as follows. $A = [2n]$, $B = [3n]$ and

$$E \subset A \times B, \quad E = \{(x + a_i, x + 2a_i) : x \in [n], i \in [R]\}.$$

G_{5n} is the union of the n matchings

$$M_x = \{(x + a_i, x + 2a_i) : i \in [R]\}.$$

We say that a subgraph $H \subset G$ is induced in G if the restriction of G to $V(H)$ equals H.

Claim. *The matchings M_x are induced in G.*

Here we sketch two different ways to see this.

Geometrically: if we had a 'cross-edge', the slopes would form an arithmetic progression.
Algebraically: $x + 2a_i = y + 2a_k$ and $x + a_j = y + a_k$ would imply $2a_i - a_j = a_k$, an arithmetic progression.

Now the main tool for proving Theorem 20 is the following result.

Theorem 21 (Induced matchings). *If G_n is the union of n induced matchings, then $e(G_n) = o(n^2)$.*

Now this indeed proves Theorem 20, since we have $Rn = e(G_{5n}) < 2\varepsilon(5n)^2 + n\varepsilon(5n)$ provided $5n \geqslant 2M(\varepsilon)/\varepsilon^2$. Hence, $R = r_3(n) < 55\varepsilon n$. $\qquad\square$

Theorem 21, in turn, is a corollary of the next lemma (used with $d = 2\varepsilon$).

Lemma 22. *Let G_n be* pure *with parameters ε, d, write $\beta = d - \varepsilon$, and assume $N \geqslant 1/(\beta\varepsilon)$ (which is certainly satisfied if $n \geqslant 2M(\varepsilon)/(\beta\varepsilon)$). If* IM *is an induced matching in G_n, then $|\mathrm{IM}| \leqslant \varepsilon n$.*

Proof. Write $U = V(\mathrm{IM})$ for the vertex set of IM, and $U_i = U \cap V_i$. Define $I = \{i : |U_i| > \varepsilon|V_i|\}$, and set $L = \cup_{i \in I} U_i$ and $S = U \setminus L$. Clearly $|S| \leqslant \varepsilon n$. Hence, if we had $|U| > 2\varepsilon n$, then we would have $|L| > |U|/2$, and thus there would exist two vertices $u, v \in L$ adjacent in IM. Let $u \in V_i$ and $v \in V_j$. We would thus have an edge between V_i and V_j, and hence, by *purity*, a density more than $\beta + \varepsilon$ between them. The sets U_i and U_j, being of size larger than εN each, would have a density more than β between them. This means more than $\beta|U_i||U_j| \geqslant \min\{|U_i|, |U_j|\}$ edges, a contradiction with IM being induced. $\qquad\square$

4. General framework for embedding H into G

A typical embedding procedure using the Regularity Lemma has several distinct phases: the last one is our main focus here.

(A) Prepare H
(B) Prepare G
(C) Assignment
(D) Making connections
(E) Piecewise embedding

Here are these five steps in more detail. We will use Alon and Yuster's proof of Theorem 8 to illustrate this structure. Since our notation H may be confused with the H there, we will use H for the whole graph to be embedded, so that in the Alon and Yuster example H is a vertex-disjoint union of copies of a fixed graph F.

(A) Prepare H. That is, chop it into (a constant number of) small pieces. Since H in Theorem 8 is a union of cn copies of F, the subdivision here is simply grouping them into clusters of εn copies.

(B) Prepare G. This has a number of steps, as follows.

(B1) Apply the Regularity Lemma to G.
(B2) Clean up the 'noise' to make it *pure* (that is, apply the degree form of the Regularity Lemma – degrees don't drop much), and define the reduced graph R.

We often need the following step.

(B3) Select a large (or perfect) matching in R (using standard matching theory)
      or, in general, a large K_r-factor (using Theorem 13).

In the example, we use $r = \chi(F)$, since every graph of chromatic number r is a subgraph of $K_r(\ell)$ for some ℓ. In fact, the union of r vertex-disjoint copies of F is a spanning subgraph of $K_r(\ell)$ with $\ell = v(F)$. Since there is a slack ε in Theorem 8, the degrees in the reduced graph R are still over $(1 - 1/r)v(R)$, so Theorem 13 can be applied.

(B4) Also, we sometimes move some vertices around to change some of the cluster sizes a little to match in size those of the small pieces of H, and/or to make some of the regular pairs super-regular.

For steps (B3) and (B4), the following trivial lemma is usually sufficient. If $\delta(G_n) \geq \left(1 - \frac{1}{r} + \varepsilon\right) n$ (where n is large), then G contains a tiling with super-regular r-cliques with only a constant number of left-over vertices. In the example, we have to make all cluster sizes divisible by $v(F)$, which can be achieved by discarding a constant number of vertices.

In the much harder cases of full-size H and with no slack in the degree conditions, we need a structural lemma that guarantees a Hajnal–Szemerédi structure in the reduced graph even under the weaker condition $\delta(G_n) \geq \left(1 - \frac{1}{r}\right) n$ *unless G_n is very special*. This is not the topic of the present paper so we only give an illustration here for the case $r = 2$ (for larger r it is too technical to spell out). A 'special' graph is an almost complete bipartite graph plus some left-over vertices; more precisely, a graph $G = (V, E)$, $|V| = n$, such that V contains two disjoint subsets A and B with

(1) $|A|, |B| > (1 - \varepsilon)n/2$
(2) $\deg(a, B) > (1 - \varepsilon)|B|$ for all $a \in A$ and the same for B.

For such special graphs, the theorems have to be proved directly.

(C) Assignment. An assignment is a map

$$\psi : V(H) \to \{V_1, \ldots, V_k\},$$

such that

$$|\psi^{-1}(V_i)| \leq |V_i| \quad \text{for all } i, \quad \text{and}$$
$$\{x, y\} \in E(H) \Rightarrow \{\psi(x), \psi(y)\} \in E(R)$$

In other words, we need to find an embedding $\varphi : V(H) \to V(G)$ of H into the closure G^c of G, where G^c is defined by replacing, for all $i < j$, the regular pair between V_i and V_j by the complete bipartite graph between V_i and V_j.

(This is the step in which we formally assign the small pieces of H to those of G.) This step is often just a simple problem about integers. In our example, we just assign vertices of the colour classes of $K_r(\ell)$ to corresponding clusters in some regular r-cliques of G.

(D) Making connections. That is, make the necessary connections *between* the (constant number of) pieces of G. This step is often easy, but sometimes tricky or even very hard.

Blissfully, H in the Alon and Yuster example is disconnected, and this step is missing. The example of embedding small powers of a Hamilton cycle is more typical. Here the small pieces are obtained by chopping the base cycle into a constant number of intervals, and there are a few edges going across these pieces. (See the remarks about band-width above, especially in Conjecture 16.)

(E) Embedding individual pieces. In principle, the whole embedding procedure is about finding an injection

$$\varphi : V(H) \to V(G)$$

such that

$$\varphi(x) \in \psi(x) \quad (\varphi \text{ is compatible with } \psi)$$
$$\{x, y\} \in E(H) \Rightarrow \{\varphi(x), \varphi(y)\} \in E(G) \quad \text{(embedding)}.$$

But we usually do it *piece by piece*, and this phase is about embedding the small parts $\tilde{H}_i$ of H into the corresponding parts $\tilde{G}_i$ of G. This is typically the hardest part, and the rest of the paper is only about embedding the small pieces $\tilde{H}$ into the small parts $\tilde{G}$.

5. The Blow-up Lemma

The classical papers using the Regularity Lemma have some unpleasant technical details that seem to recur again and again. A close inspection of those proofs makes it possible to distil the essence of those computations. Indeed, all the quoted packing theorems about bounded degree H that did not involve 'full-size packing' (meaning $|V(H)| = |V(G)|$) can be described as follows: first the Regularity Lemma is applied, then some kind of matching theorem is used for the reduced graph (this is often Theorem 13), and then the following simple lemma completes the proof.

Lemma 23 (Key Lemma). *Given $\delta, \varepsilon > 0$, a graph R, and a positive integer N, let us construct a graph G by replacing every vertex of R by N vertices, and replacing the edges of R with ε-regular pairs of density $\geq \delta + \varepsilon$ (possibly different pairs for different edges). Let H be a subgraph of $R(t)$ with h vertices and maximum degree $\Delta > 0$, and let $\varepsilon_0 = \delta^\Delta/(2 + \Delta)$. If $\varepsilon \leqslant \varepsilon_0$ and $t - 1 \leqslant \varepsilon_0 N$ then $H \subset G$. In fact,*

$$\|H \to G\| \geqslant (\varepsilon_0 N)^h.$$

Remark.   Note that $|V(R)|$ hasn't played any role here.

Using the fact that a large chunk of a regular pair is still regular (and changing ε_0), it is easy to replace the condition $H \subset R(\varepsilon_0 N)$ with the assumptions that

(*) $H \subset R((1 - \varepsilon_0)N)$
(**) every component of H is smaller than $\varepsilon_0 N$.

Typically, R is a complete graph of fixed order (e.g. a triangle).

While this is not very deep, it makes proofs short and transparent (and hence it's a great educational tool). Let us illuminate this by describing the main structure of a few classical proofs phrased in terms of the Key Lemma.

Example 1: Theorem 3.

(a) Apply the Regularity Lemma.
(b) Find a regular edge with a density greater than c.
(c) Apply the Key Lemma.

Remark.   One can find much larger regular edges (about $ne^{-1/c}$) than those provided by the Regularity Lemma.

Example 2: Theorem 4.

(a) Apply the Regularity Lemma.

(b) Two-colour the edges of the reduced graph R according to whether they represent regular pairs of density less than or more than $1/2$ (disregard the few irregular pairs). Use classical Ramsey theory to find a monochromatic r-clique in R.

(c) Apply the Key Lemma.

Example 3: Theorem 8.

(a) Apply the Regularity Lemma.

(b) Apply Theorem 13 to find a covering with regular r-cliques.

(c) Apply the Key Lemma (the union of many disjoint copies of a fixed graph is a bounded degree graph).

The only use of the Key Lemma is to make classical proofs more compact. However, if we could get rid of the strong restrictions (∗) and (∗∗), we would have a strong new tool at hand. We describe this tool – the Blow-up Lemma – later, but first here is the proof of the Key Lemma.

Proof. We prove the following more general estimate.

$$\text{If} \quad t - 1 \leqslant (\delta^\Delta - \Delta\varepsilon)N \quad \text{then} \quad \|H \to G\| > \left[(\delta^\Delta - \Delta\varepsilon)N - (t-1)\right]^h. \tag{5.1}$$

We embed the vertices $v_1, \ldots, v_h$ of H into G by picking them one by one. For each v_j not yet picked, we keep track of an ever-shrinking set C_{ij} that v_j is confined to, and we only make a final choice for the location of v_j at time j. At time 0, C_{0j} is the full N-set that v_j is *a priori* restricted to in the natural way. Hence $|C_{0j}| = N$ for all j. The algorithm at time $i \geqslant 1$ consists of two steps.

Step 1: picking v_i. We pick a vertex $v_i \in C_{i-1,i}$ such that

$$\deg_G(v_i, C_{i-1,j}) > \delta |C_{i-1,j}| \quad \text{for all} \quad j > i, \ \{v_i, v_j\} \in E(H). \tag{5.2}$$

Step 2: updating the C_js. We set, for each $j > i$,

$$C_{ij} = \begin{cases} C_{i-1,j} \cap N(v_i) & \text{if } \{v_i, v_j\} \in E(H), \\ C_{i-1,j} & \text{otherwise.} \end{cases}$$

For $i < j$, let $d_{ij} = \#\{\ell \in [i] : \{v_\ell, v_j\} \in E(H)\}$.

Fact. If $d_{ij} > 0$ then $|C_{ij}| > \delta^{d_{ij}} N$. (If $d_{ij} = 0$ then $|C_{ij}| = N$.)

Thus, for all $i < j$, $|C_{ij}| > \delta^\Delta N \geqslant \varepsilon N$, and hence, when choosing the exact location of v_i, all but at most $\Delta\varepsilon N$ vertices of $C_{i-1,i}$ satisfy (5.2). At most $t - 1$ of them have been used up before (and this is the pessimistic step here!) and consequently, we have at least

$$|C_{i-1,i}| - \Delta\varepsilon N - (t-1) > (\delta^\Delta - \Delta\varepsilon)N - (t-1)$$

free choices for v_i, proving the claim. $\qquad\square$

The interesting (and hard) problems don't have the above-mentioned two properties:

(*) small components of H

(**) slack: $|V(H)| < (1 - \varepsilon)|V(G)|$.

The following strengthening is a powerful general tool for embedding bounded degree graphs (and often trees, too).

Lemma 24 (Blow-up Lemma: short form). *Given $\delta, \varepsilon > 0$, a graph R, and a positive integer N, let us construct a graph G by replacing every vertex of R by N vertices, and replacing the edges of R with (ε, δ)-super-regular pairs. Let H be a subgraph of $R(N)$ with maximum degree $\Delta > 0$. If $\varepsilon \leqslant \varepsilon_0(\delta, \Delta)$ then $H \subset G$.*

Lemma 25 (Blow-up Lemma: full form [18, 19]). *Given a graph R of order r and positive parameters δ, Δ, there exists a positive $\varepsilon = \varepsilon(\delta, \Delta, r)$ such that the following holds. Let $n_1, n_2, \ldots, n_r$ be arbitrary positive integers and let us replace the vertices $v_1, v_2, \ldots, v_r$ of R with pairwise disjoint sets $V_1, V_2, \ldots, V_r$ of sizes $n_1, n_2, \ldots, n_r$ (blowing up). We construct two graphs on the same vertex set $V = \cup V_i$. The first graph $\mathbf{R}$ is obtained by replacing each edge $\{v_i, v_j\}$ of R with the complete bipartite graph between the corresponding vertex sets V_i and V_j. A sparser graph G is constructed by replacing each edge $\{v_i, v_j\}$ with an (ε, δ)-super-regular pair between V_i and V_j. If a graph H with $\Delta(H) \leqslant \Delta$ is embeddable into $\mathbf{R}$ then it is already embeddable into G.*

In short, *regular pairs behave like complete bipartite graphs from the point of view of bounded degree subgraphs.*

Example: the Hajnal–Szemerédi set-up. Let $\mathbf{R} = K_r$ and G a regular r-clique: that is, let the vertex set of G consist of r disjoint sets of size (arbitrary) N each, with (ε, δ)-super-regular connections between any two (where $\varepsilon \leqslant \varepsilon_0(r, \delta)$). Then G contains N vertex-disjoint K_rs. (Note that arbitrarily small (but fixed) densities δ are sufficient: we don't need large degrees as in Theorem 13.)

The Blow-up Lemma is not easy to prove. It can be proved by using a probabilistic version of the greedy algorithm used above for proving the Key Lemma, but the execution is technically complicated. For a full proof see [18]; for an algorithmic version see [19].

Remark. When using the Blow-up Lemma, we usually need the following strengthened version. Given $c > 0$, there are positive numbers $\varepsilon = \varepsilon(\delta, \Delta, r, c)$ and $\alpha = \alpha(\delta, \Delta, r, c)$, such that the Blow-up Lemma in the equal-size case (all $|V_i|$ are the same) remains true if, for every i, there are certain vertices x to be embedded into V_i whose images are *a priori* restricted to certain sets $C_x \subset V_i$ provided that

(i) each C_x within a V_i is of size at least $c|V_i|$

(ii) the number of such restrictions within a V_i is not more than $\alpha|V_i|$.

Remark. The condition that H is of bounded degree can be relaxed to p-arrangeability.

Some classical full-size embedding results easily follow from the Blow-up Lemma. Here are some recent results using (certain preliminary versions of) the Blow-up Lemma:

- a proof of Bollobás's conjecture (Conjecture 12) appears in [16] (actually, the bounded degree condition can be changed to degrees at most $cn/\log n$)
- a proof of an *approximate version* of Seymour's conjecture (Conjecture 11) appears in [17] (that of Pósa was proved earlier, in 1994, by Fan and Kierstead).

Here are some even more recent results using the Blow-up Lemma (all by Komlós, Sárközy and Szemerédi):

- a proof of Pósa's conjecture (for large n)
- a proof of Seymour's conjecture (for fixed r and large n)
- a proof of Alon and Yuster's conjecture (Conjecture 15).

We hope to get an approximate version of Conjecture 14 (Bollobás and Eldridge) and Conjecture 16 (Bollobás and Komlós).

Acknowledgement

I am grateful to the students who inspired this paper: my graph theory class Rutgers 642:587 held in the Fall of 1994. I also thank Miklós Simonovits and Béla Bollobás for their useful criticism.

References

[1] Aigner, M. and Brandt, S. (1993) Embedding arbitrary graphs of maximum degree 2. *J. London Math. Soc.* **48** 39–51.

[2] Ajtai, M., Komlós, J. and Szemerédi, E. (1992) On a conjecture of Loebl. In *Graph Theory, Combinatorics, and Applications* (Y. Alavi, and A. Schwenk, eds), *Proceedings of the Seventh Quadrennial International Conference on the Theory and Applications of Graphs* (on the occasion of Paul Erdős's 80th birthday), Kalamazoo, Michigan, pp. 1135–1146.

[3] Alon, N., Duke, R., Leffman, H., Rödl, V. and Yuster, R. (1993/4) Algorithmic aspects of the regularity lemma, *FOCS* **33** (1993) 479–481, *Journal of Algorithms* **16** (1994) 80–109.

[4] Alon, N. and Fischer, E. (1996) 2-factors in dense graphs. *Discrete Math.* **152** 13–23.

[5] Alon, N. and Yuster, R. (1992) Almost *H*-factors in dense graphs. *Graphs and Combinatorics* **8** 95–102.

[6] Alon, N. and Yuster, R. (1996) *H*-factors in dense graphs. *J. Combin. Theory Ser. B* **66** 269–282.

[7] Bollobás, B. (1978) *Extremal Graph Theory*, Academic Press, London.

[8] Chvátal, V., Rödl, V., Szemerédi, E. and Trotter Jr., W. T. (1983) The Ramsey number of a graph with bounded maximum degree. *J. Combin. Theory Ser. B* **34** 239–243.

[9] Corrádi, K. and Hajnal, A. (1963) On the maximal number of independent circuits in a graph. *Acta Math. Acad. Sci. Hungar.* **14** 423–439.

[10] Dirac, G. A. (1952) Some theorems on abstract graphs. *Proc. London Math. Soc.* **2** 68–81.

[11] Erdős, P., Füredi, Z., Loebl, M. and Sós, V. T. (1995) Discrepancy of trees. *Studia Sci. Math. Hungar.* **30** 47–57.

[12] Erdős, P. and Simonovits, M. (1966) A limit theorem in graph theory. *Studia Sci. Math. Hungar.* **1** 51–57.

[13] Erdős, P. (1963) Extremal problems in graph theory. In *Theory of Graphs and its Applications, Proceedings of the Symposium held in Smolenice in June 1963*, pp. 29–38.

[14] Erdős, P. and Stone, A. H. (1946) On the structure of linear graphs. *Bull. Amer. Math. Soc.* **52** 1089–1091.

[15] Hajnal, A. and Szemerédi, E. (1970) Proof of a conjecture of Erdős. In *Combinatorial Theory and its Applications*, Vol. II (P. Erdős, A. Rényi and V. T. Sós, eds.), Colloq. Math. Soc. J. Bolyai **4**, North-Holland, Amsterdam, pp. 601–623.

[16] Komlós, J., Sárközy, G. N. and Szemerédi, E. (1995) Proof of a packing conjecture of Bollobás. In *AMS Conference on Discrete Mathematics, DeKalb, Illinois (1993). Combinatorics, Probability and Computing* **4** 241–255.

[17] Komlós, J., Sárközy, G. N. and Szemerédi, E. On the Pósa–Seymour conjecture. Submitted to *J. Graph Theory.*

[18] Komlós, J., Sárközy, G. N. and Szemerédi, E. (1997) The Blow-up Lemma. *Combinatorica* **17** 109–123.

[19] Komlós, J., Sárközy, G. N. and Szemerédi, E. An algorithmic version of the Blow-up Lemma. To appear in *Random Structures and Algorithms.*

[20] Komlós, J. and Simonovits, M. (1996) Szemerédi's Regularity Lemma and its applications in graph theory. In *Combinatorics: Paul Erdős is Eighty*, Vol. 2 (D. Miklós, V. T. Sós and T. Szőnyi, eds), Bolyai Society Math. Studies, Keszthely, Hungary, pp. 295–352.

[21] Roth, K. F. (1954) On certain sets of integers (II). *J. London Math. Soc.* **29** 20–26.

[22] Ruzsa, I. Z. and Szemerédi, E. (1978) Triple systems with no six points carrying three triangles, In *Combinatorics (Keszthely, 1976)*, 18 Vol. II., North-Holland, Amsterdam/New York, pp. 939–945.

[23] Sauer, N. and Spencer, J. (1978) Edge disjoint placement of graphs. *J. Combin. Theory Ser. B* **25** 295–302.

[24] Seymour, P. (1974) Problem section: combinatorics. In *Proceedings of the British Combinatorial Conference 1973* (T. P. McDonough and V. C. Mavron, eds), Cambridge University Press, pp. 201–202.

[25] Simonovits, M. (1996) Paul Erdős's influence on extremal graph theory. In *The Mathematics of Paul Erdős* (R. L. Graham and J. Nešetřil, eds), Springer, Berlin, pp. 148–192.

[26] Simonovits, M. and Sós, V. T. (1991) Szemerédi's partition and quasirandomness. *Random Structures and Algorithms* **2** 1–10.

[27] Szemerédi, E. (1976) Regular partitions of graphs. In *Colloques Internationaux C.N.R.S. № 260 – Problèmes Combinatoires et Théorie des Graphes*, Orsay, pp. 399–401.

[28] Turán, P. (1941) On an extremal problem in graph theory. *Matematikai és Fizikai Lapok* **48** 436–452. In Hungarian.

[29] El-Zahar, M. H. (1984) On circuits in graphs. *Discrete Math.* **50** 227–230.

Combinatorics, Probability and Computing (1999) 8, 177–184.
© 1999 Cambridge University Press

The Homomorphism Structure
of Classes of Graphs

JAROSLAV NEŠETŘIL†

Department of Applied Mathematics, Charles University,
Malostranské nám. 25, 118 00 Prague, Czech Republic
(e-mail: nesetril@kam.ms.mff.cuni.cz)

We consider three aspects of homomorphisms of graphs and hypergraphs which are related
to the structure of colour classes: (1) density, (2) the fractal property and (3) the generation
of colour classes. In particular, we prove a density theorem for hypergraphs and show that,
for connected oriented graphs, all jumps are balanced (and give an example to show that
connectivity is needed here). We also show that a Hajós-type theorem holds for any colour
class of undirected graphs, thus providing further evidence of the well-known 'non-effective'
character of Hajós' theorem.

1. Introduction and statement of results

Graph theory receives its mathematical motivation from the two main areas of algebra
and geometry (topology, and the various notions of a graph, were present at the birth of
algebraic topology). Consequently, various operations and comparisons for graphs stress
either algebraic aspects (*e.g.* various products) or geometric aspects (*e.g.* contraction,
subdivision). It is only natural that key roles in modern graph theory are played by
(fortunate) mixtures of both approaches, as exhibited best by the various modifications of
the notion of graph minor. However, from the algebraic point of view, perhaps the most
natural graph notion is that of a homomorphism.

Given two graphs (undirected or directed) $G = (V, E)$ and $G' = (V', E')$, a homomorphism f of G into G' is a mapping $f : V \longrightarrow V'$ that satisfies the following condition:

(1) $[x, y] \in E$ implies $[f(x), f(y)] \in E'$.

In condition (1), it should be understood that the same type of edges (either undirected or
directed) are to be considered on either side of the implication. The analogous definitions
give notions of homomorphism for hypergraphs (set systems) and relational systems.

Homomorphism is an algebraic notion which, in graph theory, found its way to problems
concerning products, reconstruction and the chromatic polynomial, to name just a few.

† The author acknowledges partial support of Czech research grants GAČR 0194 and GAUK 194/1996.

Our approach here is motivated by the connection with the chromatic number expressed by the observation that, for every undirected graph G, $G \longrightarrow K_k$ if and only if $\chi(G) \leqslant k$.

Motivated by this, we call a homomorphism $G \longrightarrow H$ an H-*colouring* of G and, given H, we call the class of all graphs G that are H-colourable the *colour class determined by* H. The colour class determined by H is denoted by $\longrightarrow H$. Thus $\longrightarrow H = \{G : G \longrightarrow H\}$ is the class of all H-colourable graphs G.

The class of all colour classes determines the partially ordered class $\mathscr{C}$, ordered by inclusion. The structure of $\mathscr{C}$ is one of the subjects of this paper.

Note first that the inclusion of the colour classes $(\longrightarrow H) \subset (\longrightarrow H')$ is equivalent to $H \longrightarrow H'$. Thus the graphs H and H' determine the same colour class if and only if H and H' are homomorphism equivalent (by this we mean that both $H \longrightarrow H'$ and $H' \longrightarrow H$). This is (for finite graphs) best expressed in terms of the *core* (the core of H is the minimal subgraph of H that is a homomorphic image of H; see [7]) by saying that H and H' determine the same colour class if and only if H and H' have the same core.

The colour classes corresponding to chromatic number are the colour classes determined by complete graphs. They form a chain isomorphic to $\mathbb{N}$. The illusion of simplicity suggested by this is quickly destroyed by a moment's thought, and it appears that, contrary to this initial evidence, the structure of the class $\mathscr{C}$ is very rich. The following are extremal results in this direction; see Pultr and Trnková [13] and Welzl [14].

Theorem 1.1 (Universality Theorem). *The class $\mathscr{C}$ is universal for all partially ordered classes. Explicitly, every partially ordered class is isomorphic to an induced subclass of $\mathscr{C}$. Moreover, the class of all colour classes determined by finite graphs is universal for all countable partially ordered sets.*

Theorem 1.2 (Density Theorem). *The class of all colour classes of finite undirected graphs is (order) dense with the unique exception of the pair (K_1, K_2). Explicitly, for every pair of graphs (G_1, G_2) such that $G_1 \longrightarrow G_2$ which is not homomorphically equivalent to the pair (K_1, K_2), there exists a graph G such that G is not homomorphically equivalent to either G_1 or G_2, but $G_1 \longrightarrow G$ and $G \longrightarrow G_2$.*

Welzl's original proof of Theorem 1.2 in [14] is a difficult *ad hoc* argument. M. Perles and the author (in about 1990) independently found a much simpler and perhaps more natural proof. As neither of us published it (but several times lectured about it) the proof eventually found its way to a survey by G. Hahn and C. Tardiff [3]. Thus it is perhaps fitting to include the original proof here. This proof allows us to prove several stronger statements. These may be formulated as follows.

Theorem 1.3. *The class of all colour classes of undirected graphs is (order) dense, with the unique exception of the pair (K_1, K_2).*

Theorem 1.4. *The class of all colour classes of hypergraphs is dense.*

P_1 P_2 P_k

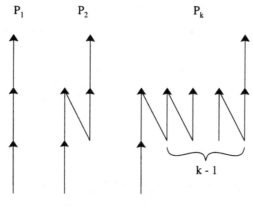

k - 1

Figure 1

Thus for hypergraphs there are no exceptional pairs, which we call *(order) jumps*. Explicitly, given a partially ordered class $\mathcal{K}$, we call a pair (A, B) of elements of $\mathcal{K}$ a *jump* if $A < B$, but there is no C with $A < C < B$.

In another direction we generalize Theorem 1.2 as follows.

Theorem 1.5. *In the class of all connected directed graphs, every jump is a pair of balanced graphs.*

(Recall that a directed graph is called *balanced* if each cycle contains an equal number of forward and backward arcs. Alternatively, a graph G is balanced if there is a homomorphism from G to a monotone path.)

The problem of characterizing all the jumps in the class of directed graphs seems to be very difficult. It is known that jumps exist. Apart from the trivial pair (single vertex, single arc) and another easy pair (single arc, monotone path of length 2), there are infinitely many jumps of the form (P_{k+1}, P_k), where the graph P_k, with $4 + 2(k - 1)$ vertices, is depicted in Figure 1. Let us call all these jumps *standard jumps*.

The existence of the jumps described above was established by the author and X. Zhu in [11], where the following (presently difficult) result was also proved.

Theorem 1.6. *The standard jumps are the only jumps in the class of all finite oriented paths.*

Even for the class of all finite directed trees, the characterization of all jumps is presently unknown. However, in view of the richness of the structure of homomorphisms between trees (see, for instance, [8] and [9]), this is perhaps hardly surprising.

However, perhaps there is more and more evidence that the standard jumps are not the only jumps, even for the class of all connected oriented graphs.

It is interesting to note that the analogue of Theorem 1.5 above does not hold in the class of all directed graphs, connected or not.

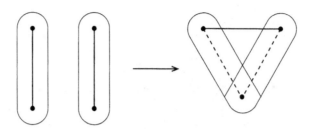

Figure 2

Proposition 1.7. *Consider the unbalanced cycle C_{2k+3} formed by an alternating path of length $2k + 1$ together with a terminal vertex joined to the first and the last vertex of this path. Let G_1 be the disconnected graph with components C_{2k+3} and P_{k+1}, and G_2 the graph with components C_{2k+3} and P_k. Then (G_1, G_2) is a jump in the class of all oriented graphs. There are infinitely many such examples, even for a fixed pair of paths.*

Let us state yet another result related to both the Density Theorem (Theorem 1.2) and the Universality Theorem (Theorem 1.1), namely that the class $\mathscr{C}$ has what we call the *fractal property*. Before defining this property, let us introduce the following definition. A mapping $F : \mathscr{C} \longrightarrow \mathscr{C}$ is said to be an *embedding* if F is 1-1 and, for any pair of colour classes A and B, $A \leqslant B$ if and only if $F(A) \leqslant F(B)$. Given a pair (A, B) of colour classes, we denote by $[A, B]$ the class of all colour classes C satisfying $A \leqslant C \leqslant B$. We call $[A, B]$ an *interval* in $\mathscr{C}$. Now we can state the fractal property of the class $\mathscr{C}$ of colour classes.

Theorem 1.8. *Let $[A, B]$ be an interval of $\mathscr{C}$ where at least one of A, B is not balanced (consequently B is not balanced). Then there is an embedding of $\mathscr{C}$ into the interval $[A, B]$.*

Thus each non-balanced interval of $\mathscr{C}$ contains a copy of the whole class $\mathscr{C}$. We shall prove this theorem elsewhere. Let us note that the fractal property is not known to be true in some cases where the Density Theorem is valid. For instance, this is the case for the class of all finite paths.

We will close with a small illustration of the usefulness of approaching problems in graph theory using homomorphisms. One of the classical results of graph theory is Hajós' theorem [4]. The result states that a graph G has chromatic number at least k if and only if G can be constructed from K_k by recursively applying the following four very simple operations: $+v$ (vertex addition), $+e$ (edge addition), $.e$ (identification of two vertices not joined by an edge; *i.e.*, contraction of a non-edge), and the so-called *Hajós join*: given two (disjoint) graphs G and G', select two edges x, y in G and x', y' in G', and build the new graph, called a Hajós join of G and G', by identifying the vertices x and x', deleting the edges x, y and x', y' and adding the new edge x', y' (see the schematic Figure 2).

While at first glance very interesting, and even surprising, Hajós' theorem is somewhat surprisingly not very useful as a practical tool, and its usefulness in chromatic graph theory is still subject to debate [10, 5]. On the other hand, it has been proved by Pitassi [12] that the minimal length of the construction of a graph by means of Hajós' theorem is related to some hard problems of proof theory. This may clarify some aspects of Hajós' theorem. Another indication of the same phenomena is perhaps the next result. Before stating it, we need some terminology.

For an undirected graph H, with independence number $k = \alpha(H)$, we define the H-*join* of graphs as follows; in fact, the definition will depend only on k.

Start with $k + 1$ points, say $0, 1, 2, \ldots, k$, and, for each unordered pair i, j ($i \neq j$), take two graphs G'_{ij} and G''_{ij}. Select edges e'_{ij} in G'_{ij} and e''_{ij} in G''_{ij} and perform a Hajós join with respect to these edges, thus obtaining a graph G_{ij}. Let a_{ij} be the identified vertex of G_{ij} (on both of the edges e'_{ij} and e''_{ij}), and b_{ij} be (say) the other vertex on e'_{ij}. Now, for each $i < j$, specify sign(i, j) to be either $+$ or $-$. Then an H-join arises from the Hajós joins G_{ij} by identifying vertex a_{ij} with point i and vertex b_{ij} with point j whenever sign$(i, j) = +$, and identifying a_{ij} with j and b_{ij} with i whenever sign$(i, j) = -$. Apart from these identifications, the graphs G_{ij} are disjoint.

Theorem 1.9. *Let H be an undirected graph. Then the class of all graphs that are not H-colourable (i.e., the class of all graphs G for which there is no homomorphism to H) can be generated from a finite set $\mathcal{H}$ of graphs by means of recursive application of the four simple operations of vertex addition, edge addition, contraction of an non-edge, and H-join.*

In essence, an H-join is a multiple Hajós join: we form $\binom{k+1}{2}$ Hajós joins, and amalgamate them according to a tournament on a given set of $k + 1$ vertices, expressed by the sign pattern.

Thus Hajós' theorem is a special case of a general theorem which holds for every colour class of undirected graphs. In view of this generality, perhaps one should not expect either too much fine structure or too many deep applications of Hajós' theorem.

The rest of the paper is organized as follows. In Section 2 we give a short proof of the Density Theorem (Theorem 1.2) together with Theorems 1.3, 1.4 and 1.5. In Section 3 we prove Theorem 1.9.

2. Density theorems: proofs

We begin with an easy proof of the Density Theorem (Theorem 1.2).

Proof of Density Theorem. (M. Perles and J. Nešetřil) Let G_1 and G_2 be given undirected graphs, let $f : G_1 \longrightarrow G_2$ be a homomorphism, and suppose there is no homomorphism $G_2 \longrightarrow G_1$. As this pair is not equivalent to the jump (K_1, K_2), at least one component of the graph G_2 has chromatic number greater than 2. Also, at least one component of G_2 fails to be G_1-colourable, and this component may be assumed to be non-bipartite; let it contain an odd cycle of length k. Now choose a graph H with the following properties: H contains no odd cycle of length k or less, and the chromatic number of H is greater than $n^{n'}$, where n and n' denote the number of vertices of the graphs G_1 and G_2 respectively. Such a graph exists by the celebrated theorem of Erdős [1]. Now let $G = G_1 \cup (H \times G_2)$. Here $\times$ denotes the direct product of two graphs and $\cup$ means the disjoint union. We shall prove that G has the desired properties. Obviously $G_1 \longrightarrow G$ and $G \longrightarrow G_2$ follows as the second projection of $H \times G_2$ is a homomorphism into G_2. On the other hand, there is no homomorphism from G_2 into G, as homomorphisms preserve odd cycles and they cannot increase the length of the shortest of them. Thus it suffices to prove that there is no homomorphism $G \longrightarrow G_1$. Let us suppose for the contradiction that there is a homomorphism $f : H \times G_2 \longrightarrow G_1$. Thus, for any vertex x of H, we have an induced mapping $f_x : V(G_2) \longrightarrow V(G_1)$ defined by $f_x(y) = f(x, y)$. (This mapping need not be a homomorphism.) As there are at most $n^{n'}$ such mappings, there are vertices x and x' forming an edge of H such that the mappings f_x and f'_x are identically equal, say to g. However in this case g is a homomorphism of G_2 into G_1, contrary to our assumption. $\qquad\square$

Proof of Theorem 1.3. We proceed exactly as in the above proof for finite graphs. The only difference is that we need a graph H without odd cycles of length k or less, with chromatic number larger than a given cardinal number. This is another folklore result of Erdős and Hajnal, which is easy to prove by considering the so-called iterated shift graphs, together with the transfinite Ramsey theorem; see [2]. $\qquad\square$

Proof of Theorem 1.5. Assume that G_1, G_2 are directed graphs with a homomorphism $G_1 \longrightarrow G_2$ but no homomorphism $G_2 \longrightarrow G_1$. We use the following observation: if G is not balanced and $G \longrightarrow G'$, then G' also fails to be balanced (this is easiest to see by using the characterization of balanced graphs as those graphs which can be coloured by monotone paths; see above). It follows that if one of the graphs G_i is unbalanced then the connected graph G_2 is also unbalanced. Let k denote the length of the shortest unbalanced cycle in G_2. Now we can proceed in a similar way to the above proof of Theorem 1.2. Let H be a directed graph with chromatic number (of its symmetrization) greater than $n^{n'}$ such that every unbalanced cycle in H has length greater than k (here we can again use shift-like graphs). Then $G = G_1 \cup (H \times G_2)$ is the desired graph. The homomorphisms $G_1 \longrightarrow G$ and $G \longrightarrow G_2$ can be obtained as before. Also, the non-existence of a homomorphism $G \longrightarrow G_1$ can be proved analogously to the above. Finally, the non-existence of the mapping $G_2 \longrightarrow G$ follows from the connectivity of G_2 together with the above observation about non-balanced graph-homomorphisms. $\qquad\square$

Proof of Theorem 1.4. In the proof we use the following hypergraph product. Given hypergraphs $H = (X, \mathcal{M})$ and $H' = (X', \mathcal{M}')$, we define their product $H \times H' = (X \times X', M \times M'; M \in \mathcal{M}, M' \in \mathcal{M}')$. It is easy to check that the projections are homomorphisms. Now, given two hypergraphs H_1 and H_2 with homomorphism $H_1 \longrightarrow H_2$ and no homomorphism $H_2 \longrightarrow H_1$, we define the desired hypergraph H as follows. First we find a hypergraph $H_0 = (Y, \mathcal{N})$ with the following properties:

(i) the chromatic number of H_0 is greater than $n^{n'}$, where n and n' are the number of points of the hypergraphs H_1 and H_2 respectively

(ii) every hyperedge $N \in \mathcal{N}$ has size greater than n' (any sufficiently large $(n'+1)$-uniform hypergraph will do).

The desired hypergraph H will be constructed as the disjoint union of H_1 and $H_0 \times H_2$. Obviously, $H_1 \longrightarrow H \longrightarrow H_2$. There is no homomorphism $H_2 \longrightarrow H$, as the set system

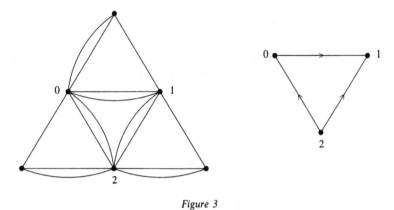

Figure 3

H_2 contains a hyperedge of size at most n', while the set system $H_0 \times H_2$ has no such edge. The fact that there is no homomorphism $H \longrightarrow H_1$ proceeds in exactly the same way as the proof of Theorem 1.2 above. (We have to use a different notion of product, but all the homomorphism properties are preserved.)   □

3. Generating colour classes: proof

Here we prove Theorem 1.9.

Proof of Theorem 1.9. Let H be a fixed undirected graph with independence number $\alpha(H) = k$. Denote by $\mathscr{H}$ the set of all inclusion-minimal graphs H' with the following two properties: the independence number of H' is at most k, and there is no homomorphism of H' into H. There is an absolute bound on the size of H' (by Ramsey's theorem, say) and thus the set $\mathscr{H}$ is a finite set. We prove that this set generates, by means of the four stated operations, exactly the colour class $\longrightarrow H$. First, we prove that the four operations preserve membership in $\longmapsto H$. However this is obvious for three of them (as in Hajós' theorem). Thus it suffices to consider the H-join only. So suppose that graphs G'_{ij}, G''_{ij}, edges e'_{ij} and e''_{ij} and sign pattern $\mathrm{sign}(i,j)$ be given (we preserve all the notation used in the definition of the H-join given earlier). Let the resulting H-join be G and suppose for the sake of contradiction that there is a homomorphism $f : G \longrightarrow H$. We consider the possible images of the amalgamation vertices $0, 1, 2, \dots, k$. Observe that, for distinct i, j, we have $f(i)$ and $f(j)$ distinct, as if $f(i) = f(j)$ then the restriction of f to the vertex set of G''_{ij} is a homomorphism from G''_{ij} to H. Thus the set $\{f(i) : i = 0, 1, \dots, k\}$ is an independent set in H, which contradicts the assumption that $\alpha(H) = k$.

   Now we prove, by a variation of the standard argument, that the class $\longmapsto H$ coincides with the class of of all constructible graphs. We proceed by double induction on the number of vertices and the number of non-edges of the given graph G. The boundary cases are easy to handle. So let us assume that G is a graph, $G \longmapsto\!\!\!\!/ \; H$ such that the graph $G + e$ can be constructed for every non-edge $e \notin G$. If $\alpha(G) \leqslant \alpha(H)$ then G contains a minimal subgraph G' with $G' \longmapsto\!\!\!\!/ \; H$. Again $\alpha(G') \leqslant \alpha(H)$ and thus $G' \in \mathscr{H}$ and thus also G is constructible. So suppose $\alpha(G) > \alpha(H) = k$. Assume without loss of generality that $\{0, 1, 2, 3, \ldots, k\}$ is an independent set in G. Observe that we may assume that for every i there exists a vertex i' such that i, i' is an edge of G. Define G'_{ij} as the graph G with the edge i, j added and the graph G''_{ij} as the graph G with the edge i, j' added. It is routine to check that G arises from these graphs by means of an H-join followed by some identification of non-edges.   □

As an example, let us consider $H = C_5$. In this case $\mathcal{H} = \{K_3\}$ and a triangle join has the form depicted in Figure 3. Except for Hajós' theorem and this particular case we do not know any other example where the colour class $\not\longmapsto H$ is generated by a single graph.

Acknowledgment

I would like to thank a referee for several remarks that improved the quality of this paper.

References

[1]   Erdős, P. (1959) Graph theory and probability. *Canad. J. Math.* **11** 34–38.
[2]   Erdős, P. and Hajnal, A. (1966) On chromatic number of graphs and set systems. *Acta Math. Acad. Sci. Hungar.* **17** 61–99.
[3]   Hahn, G. and Tardiff, C. (1997) Graph homomorphisms: structure and symmetry. In *Graph Symmetry* (H. Hahn and G. Sabidussi, eds), Kluwer, pp. 107–166.
[4]   Hajós, G. (1961) Über eine Konstruktion nicht n-färbbarer Graphen. *Wiss. Zeitschr. Martin Luther Univ. Halle-Wittenberg* **A10** 116–117.
[5]   Hanson, D., Robinson, G. C. and Toft, B. (1986) Remarks on the graph color theorem of Hajós. *Congressus Numerantium* **55** 69–76.
[6]   Hell, P. and Nešetřil, J. (1990) On the complexity of H-coloring. *J. Combin. Theory Ser. B* **48** 92–110.
[7]   Hell, P. and Nešetřil, J. (1992) The core of a graph. *Discrete Math.* **109** 117–126.
[8]   Hell, P., Nešetřil, J. and Zhu, X. (1996) Duality of graph homomorphisms. In *Combinatorics: Paul Erdős is Eighty*, Vol. 2, Bolyai Society Math. Studies, Keszthely, Hungary, pp. 271–282.
[9]   Hell, P., Nešetřil, J. and Zhu, X. (1996) Duality and polynomial testing of tree homomorphisms. *Trans. Amer. Math. Soc.* **348** 1281–1297.
[10] Mansfield, A. J. and Welsh, D. J. A. (1982) Some coloring problems and their complexity. *Ann. Discrete Math.* **13** 159–170.
[11] Nešetřil, J. and Zhu, X. (1996) Path homomorphisms. *Proc. Cambridge Phil. Soc.* **120** 207–220.
[12] Pitassi, T. (1992) The power of weak formal systems. PhD thesis, University of Toronto.
[13] Pultr, A. and Trnková, V. (1990) *Combinatorial, Algebraic and Topological Representations of Groups, Semigroups and Categories*, North Holland.
[14] Welzl, E. (1982) Color families are dense. *Theoret. Comput. Sci.* **17** 29–41.

Combinatorics, Probability and Computing (1999) 8, 185–192.
© 1999 Cambridge University Press

Problem Collection
of the DIMANET Mátraháza Workshop, 22–28 October 1995

1. Noga Alon

Paul Erdős [2] conjectured in 1979 that, if in a graph on n vertices any set of $\lfloor \sqrt{n} \rfloor$ vertices contains at least one edge, then there is a set of $\lfloor \sqrt{n} \rfloor$ vertices that contains $\Omega(\sqrt{n} \log n)$ edges. As observed by Erdős, this result, if true, is tight. During the workshop, and after discussions with various participants including Cameron, Erdős, Gunderson and Krivelevich, we found a proof of this conjecture, combining some probabilistic arguments with the main result of [1] (see also [3]). Hopefully this will appear in a forthcoming paper, where we also plan to include a simple proof of an extension of the main result of [1].

The following more general Ramsey-type problem is still open.

Conjecture. There exists an absolute positive constant c so that the following holds. Let G be a graph on n vertices, let m be an integer satisfying $\log n \leqslant m \leqslant n$ and suppose that every set of m vertices in G contains at least one edge. Then there is some set of m vertices of G containing at least $cm \log(n/m)$ edges.

If true the above conjecture is tight up to the constant c. We can prove it for, say, $n^{0.01} \leqslant m \leqslant n$, and it is easy to show it for $m = O(\log n)$. For all other values of m we can only prove the weaker result obtained from the above conjecture by replacing the $\log(n/m)$ term by

$$\max\{\log m, \log n / \log \log n, \log n / \log(m / \log n)\}.$$

References

[1] Ajtai, M., Komlós, J. and Szemerédi, E. (1980) A note on Ramsey numbers. *J. Combin. Theory Ser. A* **29** 354–360.
[2] Erdős, P. (1979) Some old and new problems in various branches of Combinatorics. In *Proc. 10th Southeastern Conf. on Combinatorics, Graph Theory and Computing, Boca Raton 1979*, Vol I, *Congressus Numerantium* **23** 19–37.
[3] Shearer, J. B. (1983) A note on the independence number of triangle-free graphs. *Discrete Math.* **46** 83–87.

2. Peter J. Cameron

2.1. Sum-free sets

The questions below are posed jointly with Paul Erdős. A set S of natural numbers is sum-free if it contains no x, y, z with $x + y = z$.

(a) An old problem to set the scene: how many sum-free subsets of $\{1, \ldots, n\}$ are there? It is known that there are at most $2^{(1/2+\epsilon)n}$, and at least $c \cdot 2^{n/2}$, such sets; it is conjectured that the lower bound is the truth. The Hausdorff dimension of the set of sum-free sets is $1/2$.

(b) How many sum-free subsets of $\{1, \ldots, n\}$ are maximal with respect to inclusion? There exist at least $2^{\lfloor n/4 \rfloor}$ maximal sets. For let m be either n or $n - 1$, whichever is even; now consider the set S containing m and one of each pair $x, m - x$ for odd $x < m/2$. No further odd numbers less than m can be added to these sets, so they lie in distinct maximal sets.

   We cannot prove a non-trivial upper bound. However, in attempting to show that the number is $o(2^{n/2})$, it would suffice to consider maximal sets S with $\min(S) = o(n)$ and $(1/10)n \leqslant |S| \leqslant (2/5)n$. (The upper bound depends on a theorem on the structure of large sum-free sets by Deshouillers, Freiman, Sós and Temkin.)

(c) What is the size $g(n, p)$ of the largest sum-free subset of $\{1, \ldots, n\}$ whose least element is p? We can prove an upper bound as follows:

$$g(n, p) \leqslant \begin{cases} n - p + 1 & \text{if } p \geqslant n/2, \\ p & \text{if } n/3 \leqslant p < n/2, \\ \lfloor (n - p)/2 \rfloor + 1 & \text{if } p < n/3. \end{cases}$$

This bound is attained if either $p \geqslant (n + 2)/5$ or p is odd. The exact value for even $p < (n + 2)/5$ is unknown.

(d) Let α be irrational, and let

$$S_\alpha = \{k : 1/3 < \{k\alpha\} < 2/3\}.$$

Then S_α is a maximal sum-free set. We have discovered various properties of these sets. For example, if α does not have good rational approximations, then the longest arithmetic progression contained in $S_\alpha(n) = S_\alpha \cap \{1, \ldots, n\}$ has size $O(\sqrt{n})$. We have counted the number of distinct sets $S_\alpha(n)$ exactly (for fixed n); it is asymptotically $(27/8\pi^2)n^2$. We have also constructed various generalisations of the sets S_α.

2.2. The probability of connectedness

In my lecture, I asked the following question. Let $\mathscr{X}(H)$ denote the class of graphs for which no induced subgraph is isomorphic to H. For which connected graphs H is it the case that $\text{Prob}(G \text{ is connected} \mid G \in \mathscr{X}(H))$ tends to a limit p with $0 < p < 1$? I subsequently showed that the only connected graph with this property is the path of length 3 (for which the probability is $1/2$). If we do not demand that H be connected, then the problem is open, and different techniques are required.

2.3. Paths in vertex-transitive graphs

The following question is due to D. Sotteau. Does every vertex-transitive graph G have a collection $\mathscr{P}$ of directed paths such that

(a) for any two vertices $x \neq y$, there is a unique path $P \in \mathscr{P}$ from x to y,

(b) each vertex lies on a constant number of paths in $\mathscr{P}$?

I cannot answer this question. However, I can make the following very small contribution. There is a collection of paths satisfying (a) above and also

(b') each vertex is the second-last vertex in a constant number of paths in $\mathscr{P}$,

(c) $\mathscr{P}$ contains all initial segments of all of its paths.

2.4. Orders of subgroups

The following question is due to P. Erdős. For a positive integer n, let $\phi(G)$ be the number of orders of subgroups of the group G of order n; let $f(n)$ be the minimum of $\phi(G)$ over all groups of order n. Then $f(n) \leqslant d(n)$, by Lagrange's theorem; but how small can $f(n)$ be? I could construct examples with $f(n) = d(n)^c$, where $c = \log 5 / \log 6$. I do not know whether this is close to the best possible or not.

3. Reinhard Diestel

Does an infinite graph contain a topologically end-faithful forest if its end space is metrizable?

Note that, if true, this would characterize the graphs admitting such a forest in the simplest possible way – the converse of the above statement is trivially true, since all forests clearly have a metrizable end space. (Add a root, and give two distinct ends distance $1/n$ if their unique rays from the root split after n vertices.)

Precise definitions and more references can be found in my brief survey [4] on ends of graphs, especially in Section 5. Here we confine ourselves to giving somewhat informal definitions.

An end is an equivalence class of rays (one-way infinite paths), two rays being equivalent if their tails cannot be separated by a finite set of vertices. The set of ends of a graph carries a topology: when we delete a finite set of vertices and consider a component of the remaining graph, then the set of ends 'contained' in this component is a basic open set. (An end is taken to be contained in such a component if and only if one – and hence each – of its rays has a tail in it.) A forest contained in a graph has its own ends, but each end of the forest is a subset of an end of the graph. The forest is called topologically end-faithful in the graph if this map is a homeomorphism. (The map is clearly always continuous, but it need be neither injective nor surjective, nor is its inverse (if it exists) necessarily continuous.)

References

[4] Diestel, R. (1992) The end structure of a graph: recent results and open problems. *Discrete Math.* **100** 313–327.

4. György Elekes

In a triple system, six triples will be called a *tic-tac-toe* if they intersect each other like the rows and columns of a (3×3) tic-tac-toe board.

Let $\mathcal{T}_n$ be a triple system on n points such that

(1) any two triples have at most one point in common,
(2) no six triples form a tic-tac-toe.

Is it true that $\mathcal{T}_n$ has $o(n^2)$ triples?

The problem can be reformulated in the following way in terms of graphs. For fixed $c > 0$, suppose that the edges of a graph G with n vertices and cn^2 edges are well-coloured using n colours (*i.e.*, no two edges of the same colour are incident upon a common vertex). If n is sufficiently large, must G contain a six-cycle with opposite edges having the same colour?

5. David Gunderson

The problem below arose in joint work with Deuber, Erdős, Kostochka and Meyer.

A Δ-*system* is a set-system of the form $\{A \cup A_i : 1 \leqslant i \leqslant r\}$, where the A_is are disjoint from each other and from the set A. Let $f(n,r)$ be the smallest number such that any n-uniform set-system with $f(n,r)$ elements contains an r-element Δ-system. Erdős and Rado [6] conjectured that, for every r, there exists a constant C_r such that $f(n,r) \leqslant C_r^n$.

Similarly, we can define $F(n,r)$ as the smallest number such that any, not necessarily uniform, set-system of $[n]$ with $F(n,r)$ elements contains an r-element Δ-system. Erdős and Szemerédi [7] observed that the probabilistic method yields that, for each $r \geqslant 3$, there exists a constant C_r so that $F(n,r) > (1 + C_r)^n$.

We observed in [5] that

$$2^{n(1 - \frac{\log \log r}{2r}) - O(\frac{1}{r})} \leqslant F(n,r) < 2^{n - \sqrt{\frac{n \log \log n}{\log \log \log n}}}$$

and, if the Erdős–Rado conjecture is true, that is, if there exists a C so that $f(k,3) \leqslant C^k$, then

$$F(n,3) < 2^{n(1 - 0.65/C)}.$$

Note that the right-hand side is of the form $(2 - \epsilon)^n$.

In [5] we also gave a construction showing that $1.551^{n-2} < F(n,3)$. Is it true that there exists $\epsilon > 0$ so that $F(n,3) \leqslant (2 - \epsilon)^n$? Frankl and Rödl [8] showed that such a bound holds for *weak* Δ-systems, that is, set systems whose sets have pairwise intersections the same size.

References

[5] Deuber, W., Erdős, P., Gunderson, D., Kostochka, A. and Meyer, A. (1997) Intersection statements for systems of sets. *J. Combin. Theory Ser. A* **79** 118–132.

[6] Erdős, P. and Rado, R. (1960) Intersection theorems for systems of sets. *J. London Math. Soc.* **35** 85–90.

[7] Erdős, P. and Szemerédi, E. (1978) Combinatorial properties of systems of sets. *J. Combin. Theory Ser. A* **24** 308–313.

[8] Frankl, P. and Rödl, V. (1987) Forbidden intersections. *Trans. Amer. Math. Soc.* **300** 259–286.

6. András Gyárfás

Assume that K_{r^2+1} is written as the edge-disjoint union of r graphs.

I conjecture that, for $r \geqslant 3$, the union of some $r - 1$ of these graphs contains a K_{r+1}.

The conjecture is true for $r = 3$ and 4. For $r = 2$ the statement is not true since K_5 is the union of two pentagons. The statement is not true for K_{r^2}, at least for prime power r, since the affine plane of order r has a natural decomposition.

7. Svante Janson

Several people have pointed out analogies between problems for graphs and for arithmetic sequences. Thus, inspired by the problem concerning 'proportional graphs' (see for example [9, 10]), I suggest the following problem.

Construct (or show the existence of) a finite 0–1 sequence such that, among all subsequences of three elements indexed by an arithmetic progression, exactly 1/8 are of each of the types 000, 001, 010, 011, 100, 101, 110, 111. (Note that these are the expected numbers for a random sequence: the problem is to find a sequence that exactly yields these expected values.)

Many variations are possible: for example, one may consider different proportions, more than two different numbers, longer subsequences, or cyclical sequences.

Small examples may be interesting, but the existence of arbitrarily long such sequences is probably more interesting.

References
[9] Janson, S. and Kratochvil, J. (1991) Proportional graphs. *Rand. Struct. Alg.* **2** 209–224.
[10] Janson, S. and Spencer, J. (1992) Probabilistic construction of proportional graphs. *Rand. Struct. Alg.* **3** 127–137.

8. Zoltán Király

The following problem is a generalization of a geometric problem I heard from Paul Erdős.

Define the distance of two sets A and B as

$$d(A, B) = \sqrt{|A \setminus B|^2 + |B \setminus A|^2}.$$

Prove that, if $\mathscr{A}$ and $\mathscr{B}$ are families of m sets on the same ground-set, then

$$\sum_{\substack{A \in \mathscr{A} \\ A' \in \mathscr{A}}} d(A, A') + \sum_{\substack{B \in \mathscr{B} \\ B' \in \mathscr{B}}} d(B, B') \leqslant \sum_{A \in \mathscr{A}} \sum_{B \in \mathscr{B}} d(A, B).$$

9. Christian Mauduit

Let A be a finite alphabet, and take $w = w_0 w_1 \ldots w_n \ldots \in A^{\mathbb{N}}$. Let $L(w) = \{\mu \in A^* \mid \exists w' \in A^*, \exists w'' \in A^{\mathbb{N}} : w = w' \mu w''\}$.

Let $p(n) = |L(w) \cap A^n|$ be the complexity function of w. It is obvious that p is a monotone increasing function and $1 \leqslant p(n) \leqslant |A|^n$. It is also easy to see that, if $p(n_0) \leqslant n_0$ for some n_0, then $p(n)$ is bounded.

The problem is to characterize the complexity functions.

10. Peter Mihók

Let us denote by I the set of all mutually nonisomorphic finite undirected simple graphs. If P is a nonempty subset of I, then P will also denote the graph property of being a member of the set P.

Let $\Delta(G)$ denote the maximum degree of a graph G, and $\tau(G)$ the length of a longest path, and consider the following properties:

$$S_k = \{G \in I : \Delta(G) \leqslant k\} \quad \text{and} \quad Q_k = \{G \in I : \tau(G) \leqslant k\}.$$

For properties P, Q of graphs, define:

$$P \circ Q = \{G \in I : V(G) = V_1 \cup V_2, \ G[V_1] \in P, \ G[V_2] \in Q\}.$$

With this notation, the well-known Lovász decomposition theorem [11] states that, for all integers $p, q \geqslant 0$, $S_{p+q+1} \subset S_p \circ S_q$.

The corresponding question for the property Q_k is still open.

Problem 1. *Is it true that $Q_{p+q+1} \subset Q_p \circ Q_k$ for all $p, q \geqslant 0$?*

If true, this would be related to the next question.

Problem 2. *Let H_v be a graph with root v. Call a set $S \subset V(G)$ an H_v-kernel of G if*

(i) *$G[S]$ does not contain any subgraph isomorphic to H_v,*
(ii) *for all $v' \in V(G) - S$, $G[S \cup \{v'\}]$ contains an isomorphic copy H of H_v, where v' is the root of H.*

For which H_v is it true that every graph G has an H_v-kernel?

This is not the case for every graph H_v: the smallest known counterexample has five vertices.

However, there are many graphs H_v such that every graph G has an H_v-kernel. For example, Lawrence in 1978 proved that if H_v is a star with the central vertex as the root, then every graph G has an H_v-kernel. As far as we know, for $H_v = P_n$, a path on n vertices, with the root v being one of the vertices of degree 1, an affirmative answer to the second problem is known only for $n \leqslant 6$.

The problems above seem much harder for directed graphs. However, Péter Hajnal asked about the following beautiful special case of the directed version of Problem 1.

Let G be a directed graph. Is it true that there exists an independent set of vertices that covers all longest directed paths?

References
[11] Lovász, L. (1966) On decomposition of graphs. *Studia Sci. Math. Hungar.* 237–238.

11. Imre Ruzsa

Is there a Sidon set $A \subset [1, 2, ..., N]$ such that $|A| > N^{1/2-\varepsilon}$ and $|A \cap [1, N^{2/3}]| > N^{1/3-\varepsilon}$?

12. András Sárközy

For natural numbers k and N, let $f_k(N)$ be the maximum cardinality of a subset A of $[N]$ such that no product $a_1 a_2 \ldots a_k$ of k distinct elements of A is a perfect square.

We know that $f_4(N) = (1 + o(1))\pi(N)$, $f_6(N) = (1 + o(1))\pi(N)$, and $f_3(N) = (1 + o(1))N$. The determination of f_k seems to be much more difficult for odd values of k. We do not even know the approximate behaviour of $f_5(N)$.

13. Dominique Sotteau

The following conjecture is due to Heydemann, Meyer and Sotteau.

In any vertex-transitive graph of order n, there exists a set of $n(n-1)$ shortest paths between all the pairs of vertices (*i.e.*, one path from any x to any y) such that the same number of paths go through each vertex.

This is known to be true for Cayley graphs ([12]).

Note that the path from x to y is not necessarily the same as the one from y to x. In fact, it would be false if this condition were required, as can be observed easily by taking for G a 4-cycle.

References
[12] Heydemann, M. C., Meyer, J. C. and Sotteau, D. (1989) On forwarding indices of networks. *Discrete Appl. Math.* **23** 103–123.

14. László Székely

De Caen and Székely proved that in the incidence bipartite graph of n points and m lines in the Euclidean plane the number of 3-paths is $O(nm)$. By Atkinson's inequality this easily implies the Szemerédi–Trotter bound for the number of edges of the incidence graph.

We have a stronger conjecture, namely, that the number of 6-cycles (*i.e.*, of triangles for geometers) is also $O(nm)$. Although the simplicity of the conjectured bound cries out for induction, such a proof has so far eluded us.

15. Pavel Valtr and Jaroslav Nešetřil

Fürstenberg, Katznelson and Weiss [15] proved the following two results.

Theorem 1. *For any k-colouring of the plane there is an l_0 such that, for any real $l \geq l_0$, one colour class contains two points at distance l.*

Theorem 2. *For any k-colouring of the plane and for any 3-point configuration T, there is a number l_0 such that, for any $\varepsilon > 0$ and for any real $l \geq l_0$, one colour class contains a monochromatic ε-approximation of a set congruent to lT.*

Theorems 1 and 2 are proved in [15] by means of ergodic theory. Theorem 1 has also been proved by other ᴊᴇthods by Bourgain [13] and by Falconer and Marstrand [14]. Bourgain [13] also gave a refinement of Theorem 1 in higher dimensions.

Two configurations (*i.e.*, finite planar point sets) are said to be of the same *order type* if there is a bijection between them which preserves orientations of triples of points. We say that a configuration B is *induced* in a configuration A, if $B \subseteq A$ and conv(B)$\cap A = B$.

In the spirit of the above results of Fürstenberg, Katznelson and Weiss, and of other Ramsey-type results in the plane, we have proved [16] the following related result.

Theorem 3. *For any configuration P and for any integer $k \geqslant 2$, there is a configuration $R = R(P,k)$ such that, for any partition of R into k colour classes, there is a subset $\tilde{P}$ of R with the following three properties:*

   (i) *$\tilde{P}$ is monochromatic (i.e., it is a subset of one of the colour classes),*
  (ii) *$\tilde{P}$ and P are of the same order type,*
 (iii) *$\tilde{P}$ is induced in R.*

Theorem 3 has no higher-order analogues even if we drop condition (iii).

Theorem 4. *For every $p \geqslant 2$, there exists a finite planar point set $P(p)$ in general position with the following property. There exists a partition $\binom{\mathbb{R}^2}{p} = C_1 \cup C_2$ of all p-element subsets of the plane into two colour classes such that no monochromatic subset of $\mathbb{R}^2$ is of the same order type as $P(p)$.*

However, we have no bound on how large the set $P(p)$ in Theorem 4 is. In [16] we gave a relatively complicated argument that the size of $P(2)$ must be at least 5, but we don't even know if this is also true if we consider a higher-order analogue of Theorem 3 with condition (iii). Thus we pose the following problem.

Is it true that, for any configuration P of at most four points in general position and for any integer $k \geqslant 2$, there is a configuration $R = R(P,k)$ such that, for any partition of $\binom{R}{2}$ into k colour classes, there is a subset $\tilde{P}$ of R with the following three properties:

   (i) *$\binom{\tilde{P}}{2}$ is monochromatic (i.e., it is a subset of one of the colour classes),*
  (ii) *$\tilde{P}$ and P are of the same order type,*
 (iii) *$\tilde{P}$ is induced in R.*

References

[13] Bourgain, J. (1986) A Szemerédi type theorem for sets of positive density in $\mathbb{R}^k$. *Israel J. Math.* **54** 307–316.

[14] Falconer, K. J. and Marstrand, J. M. (1986) Plane sets with positive density at infinity contain all large distances. *Bull. London Math. Soc.* **18** 471–474.

[15] Fürstenberg, H., Katznelson, Y. and Weiss, B. (1990) Ergodic theory and configurations in sets of positive density. In *Mathematics of Ramsey theory* (J. Nešetřil and V. Rödl, eds), Springer, pp. 184–198.

[16] Nešetřil, J. and Valtr, P. (1994) A Ramsey-type theorem in the plane. *Combinatorics, Probability and Computing* **3** 127–135.